CIRCULARITY AND RESOURCE-BASED

DISPUTE

MANAGEMENT SYSTEM

A Guide for Environmental Dispute Managers

By

Larry D. Long

E-Book: 978-1-291-81787-4

Paperback: 978-1-291-81786-7

Hardcover: 978-1-291-81785-0

Foreword

Rachel Carson once said "Man's attitude toward nature is today critically important simply because we have now acquired a fateful power to alter and destroy nature. But man is a part of nature, and his war against nature is inevitably a war against himself."

Environmental mediation involves the interplay between the environment, resources, and conflict, which is as evident today as the established links among human rights, democracy, and peace. Environmental conflicts draw upon expertise from various fields, including science, law, economics, and political ideology, all of which converge to form objective criteria for achieving sustainable resolutions.

The Circularity and Resource-Based (CRB) Dispute Management System is designed to integrate natural resources with cultural diversity into negotiation and mediation, using systematic management tools. This approach enables disputants to objectively develop and choose resolutions that benefit all parties while promoting environmental conservation. When applied correctly, this process can help resolve long-standing resource and environmental disputes while safeguarding resources for future generations.

When appropriately applied, the CRB system can improve environmental negotiations by redirecting parties' interests to core natural resources or environmental issues. It integrates and analyzes related symptomatic concerns early in the process, developing "Objective Criteria" alongside the core issues of the main dispute.

Acknowledgements

This book draws on over thirty years of my research in hydrology, environmental and natural resource law, and dispute resolution, fields in which I hold graduate degrees.

About The Author

Larry Long serves as CEO of International Dispute Management LLC and brings four decades of environmental expertise to his role. His experience includes twenty-one years with the U.S. Environmental Protection Agency in regulatory and enforcement, with specialized knowledge in wetlands, the National Environmental Protection Act (NEPA), and mining. Mr. Long holds a Bachelor of Science in Biology/Biochemistry, a Master of Science in Watershed Hydrology, a Master of Science in Dispute Management, and an Executive Juris Doctor degree in Law.

Long is a proven leader skilled in mediation and management, applying dispute resolution to environmental disputes that foster collaboration among diverse stakeholders. He has recognized technical expertise in hydrology, water resources, mining, environmental analysis, and transportation, with multiple awards and publications. He is the author of the book Resource-Based Dispute Management: A Guide for the Environmental Dispute Manager, May2015, as well as a Master's Thesis, the Transport of Cryptosporidium through the Perched Zone of a Wastewater Effluent Treatment Basin, March 1999, and a journal article, The Introduction to the Conceptual Resource-Based Mediation System, Duquesne Law Journal Joule, January 2014.

@internationaldisputemgm8488

An Introduction to CRB

Dispute Management System

Table of Contents

Chapter 1
Introduction

Historical, international, and national conflicts have served as significant influences on social and political dynamics. Many of these factors evolve into elevated and often violent conflicts, including war. Elevated or protracted disputes have enormous potential to ignite wars that are passed from generation to generation and from country to country. The need to resolve disputes early is therefore in everyone's best interest. Following World War I, the League of Nations established procedures to address prolonged disputes through peaceful means rather than through violence. Yet current news reports indicate that more of these disputes are now being managed through violence and armed conflict.

Environmental disputes are no different in that their root causes are typically based on economic, legal, social, and political issues. I regard extended disputes as opportunities to achieve resolution, regardless of the challenges they may present. It is important to recognize, however, that while opportunities may arise, achieving resolution is not always immediate, and long-term success can require decades of sustained effort.

Incidents of violence stemming directly from environmental disputes are rare; however, conflicts over natural resources have demonstrated varied outcomes. Environmental factors such as drought may contribute to tensions that take years to escalate into open conflict. In most cases, legal proceedings are initiated first, as is commonly observed with water-related disputes. In recent decades, the process of resolving environmental and natural resource disputes has gained prominence due to factors such as drought, global climate change, and growing demand for mineral resources. Another reason environmental disputes have been brought to the forefront is the occurrence of catastrophes, both natural and human-caused, as well as increasing limitations on available resources. Throughout human history, the resolution of disputes has increasingly shifted toward methods that prioritize mutual interests rather than conflict. Violence rarely yields constructive outcomes, and war typically results in significant challenges related to governance and recovery, offering limited benefits to any party involved.

The processes for managing disputes emerged from the establishment of rules, regulations, and laws, all of which were designed to provide more equitable resolutions. These laws, rules, and regulations evolved over time to meet the needs of society at particular historical moments. The equitable distribution envisioned by the law is often translated into a monetary settlement. However, monetary settlements involving natural resources are frequently complex, as it can be challenging to accurately assess the true value of these resources.

We are often told that we live by a set of rules, including the Golden Rule, which was, by most accounts, as early as 551 B.C. The Golden Rule, a Confucianism maxim, defines a two-way relationship of equal treatment[1]. The management of natural resources directly translates into influence. It is important to recognize that influence serves as a key factor in resolving disputes. In contemporary discourse, the commonly understood meaning of the Golden Rule has evolved, too: "He who possesses wealth establishes the rules," highlighting how economic influence is often perceived as a source of power in modern society. Alternatively, we commonly refer to the *phrase only*

[1] The Golden Rules crosses most every culture since recorded time. No one is truly sure where it started but many take claim to it origin.

the strong survive[2], indicating a win-lose or Zero-Sum type resolution, with success going to the victor. Our perspectives on how we apply these maxims toward environmental issues reflect some of the power dynamics of environmental disputes.

How we deal with environmental disputes demonstrates how we as people have evolved over time, and how we as a community, a country, or as humans globally have evolved to resolve our issues through a peaceful and equitable process. It is said that the Romans thought that to live in peace, one must prepare for war[3]. This type of self-preservation did not necessarily serve the Roman Empire well, but it does reflect the cultural power structures of that time in history, as well as those of today. This saying may seem a little drastic when applied to environmental disputes, but it captures the all-or-nothing mentality that often permeates such conflicts on both sides. The same applies to environmentalism more broadly: we all want to live in an environment that is clean and free of pollutants that contribute to health problems, but at what cost? We must therefore examine the true costs, both positive and negative, associated with maintaining an environment that supports good health, along with the arguments surrounding those costs.

The argument that environmental regulations are too costly hinges on concepts similar to those applied when preparing for war. Politicians often elevate dogma to capture public attention by claiming they are "waging war" on an issue. That is why we have so many issue-based wars, such as the war on poverty, the war on crime, the war on drugs, and so on. Many governments and individuals still operate

---

[2] The phrase "only the strong survive" is quoted from Herbert Spencer after he reportedly read Darwin's 5th edition Origins of Species, 1869.

[3] Publius Flavius *Vegetius* a Roman Military expert in the 4th century A.D.

within this mindset. However, this dogma-driven mentality only builds mental, political, and social walls that hinder peaceful solutions.

Building walls in preparation for war, whether physical or metaphorical, has not proven to be a lasting solution for sovereign states or for environmental issues. The Israeli–Palestinian conflict provides a clear example: building a wall has not stopped the violence or resolved the underlying problems. This level of thought reflects a mindset with inherent weaknesses, one in which a nation is always preparing for war and rarely has the time or stability to live in peace. Residing in the United States, a nation recognized for possessing one of the largest militaries globally and allocating substantial resources to support it, makes this issue particularly pertinent because of the significant effects on daily life.

We all want to live in peace, but again, at what cost? If we look around the world, the Roman mindset is clearly represented in almost everything we do and is evident in most countries' foreign policies, as well as in the way we view our global and national natural resources. In regions where prolonged conflicts over sovereignty have occurred, the Roman approach may be considered necessary or viewed as an expected consequence. This perspective enables the use of fear-based political tactics, sometimes resulting in circular reasoning. The application of fear, whether in national or environmental debates, is often used as a persuasive and influential tool.

This mentality is also obvious in environmental and natural resource disputes, where the interest-based system is used to resolve the issues and provide a resolution based on perceived interest masquerading as a need[4]. To reverse this type of attitude is extremely

---

[4] A need is defined as a fundamental requirement for human survival and well-being, including essential components such as food, water, shelter, and healthcare. In contrast to wants, which are discretionary and based on individual preferences, needs are non-

difficult because everyone wants their slice of the global resource pie[5]. So how might a practitioner propose to resolve an environmental or natural resource dispute in which the disputants have a history of self-preservation and no clear vision for the future conservation of the resource?

Natural resources are often viewed as commodities and therefore as assets. Using the circularity concept tool in dispute resolution can address the interests of involved parties while also considering the management of natural resources. This approach is not suitable for all resources due to limited life cycles, but when a resource can be recycled multiple times, it may offer a potential solution. These assets provide political and economic advantages that translate into power for those who can control and manipulate the market. Therefore, regardless of the environmental and human costs associated with particular resources, market value frequently remains the primary focus. All nations aim to safeguard their assets, and political leaders often emphasize the importance of protecting national interests, which may, at times, involve securing access to natural resources located in other regions of the world.

As we consider current global political and economic uncertainties, it becomes clear that the past decade or two has taken a toll on both renewable and non-renewable natural resources. The fact that our national and global reserves are dwindling at a much greater rate than expected will only further complicate global economies.

---

discretionary and vital for sustaining basic life functions. Failure to meet these needs may result in adverse outcomes, affecting both personal health and the stability of society.

[5] Global- is defined as not only issues all over the planet but also include all the issues associated with individual projects.

An example of how this global chess match is being played can be seen in the strategic positioning of certain countries regarding rare earth metals or natural gas. The greatest known reserves of rare earth metals are found in China, which is believed to hold as much as 70 percent of known deposits. However, as the world searches for new sources of critical rare earth metals (CREM), new deposits are being identified in various regions. By examining the life cycle of CREM, we also discover that they can be recovered by recycling electronics and other materials. CREM plays a critical role in the manufacture of advanced batteries, mobile devices, computers, communication equipment, and, notably, defense applications such as smart weaponry. Both the United States and the European Union have initiated efforts to develop technological alternatives to CREM, aiming to shift global market dynamics for these materials.

Another example is Russia's use of natural gas as a means to pressure Ukraine and other former Soviet states into returning to a sphere of influence resembling the former Soviet structure. This strategy has led to violent incidents on several occasions over the past year. Such an aggressive approach demonstrates how a resource, natural gas, can be used by one country to position itself in power over several smaller nations for economic gain. Power, however, is not the only factor; natural and human-caused conditions also contribute to resource conflicts. This mentality is reflected as well in the broader dynamics of global climate change.

The Klamath River dispute, involving the removal of four hydroelectric dams, provides another instructive example. The project was initiated to restore the river's natural flow, which had been obstructed for more than a century, significantly affecting fish populations and water quality. The dispute illustrates the complex power relationships among water resource management, environmental conservation, and tribal rights. The recent removal of the dams marks a considerable advancement in enhancing salmon habitats and revitalizing the river's ecosystem. As the largest dam-removal project in U.S. history, it represents years of advocacy by Indigenous tribes and environmental supporters seeking to reconnect with ancestral waterways and restore natural river dynamics. The agreement reached between agricultural stakeholders and tribal communities is intended to improve water-supply reliability for farming while simultaneously advancing ecological restoration. This notable accord highlights the crucial role of collaborative engagement among diverse parties in promoting the long-term sustainability of the Klamath River.

In the natural resources arena, conflicts associated with natural disasters may result from direct or indirect consequences of changing weather patterns or may be intensified by global climate change, which has increased demand for certain resources. The crisis in Somalia was first brought to the world's attention as a drought that triggered a series of social and economic problems for the local population and quickly spread to neighboring communities. Drought is a normal and predictable part of weather cycles. However, the widespread use of fossil fuels has increased global temperatures. When drought conditions extend beyond predictable patterns and disrupt growing and grazing cycles, they have the potential to escalate into more than just a local problem. Without a greater understanding of global climate change, we may find ourselves in a world with a disproportionate distribution of natural resources. It is easy to see how global conflict could arise, and continue to increase, based on resource shortages driven by the altered weather patterns brought about by global climate change.

If we continue this course of self-interest with disregard for evidence of changing weather patterns, focusing only on short-term economic factors, global conflicts will have the potential of becoming a major part of our history. Many believe that global climate change has contributed to the "Death of Environmentalism,"[6] as some have termed it, or would it be more productive to focus on climate change issues that really matter?[7]

The predominant rationale for prioritizing economic markets over environmental concerns is the emphasis on profits, which are sold to the general population as growth and job creation. The industry has effectively persuaded workers that environmental regulations are the cause of job loss. However, this approach exacerbates tensions without addressing the underlying issues that impact public health. On the contrary, numerous regulatory environmental incentive programs contribute to job creation. As evidenced during the recent recession, the environmental clean-up industry supports employment in various roles, including consultants, scientists, engineers, regulators, and construction professionals. Disputes have shaped human history, often resulting in violence. After World War 1, the Treaty of Versailles[8] emerged to resolve conflicts peacefully. While early philosophies like Confucianism promoted mutual respect, modern interpretations often favor power dynamics, with winners taking control. How societies manage disputes reflects their evolution toward fairness and peace, yet ongoing global conflicts challenge the notion of real progress. The imperialist notion of Manifest Destiny, which asserted that early settlers were destined to expand westward across North America, fostered an approach focused on the consumption of natural

---

[6] Shellenberger, M. and Nordhaus, T., (2004) The Death of Environmentalism: Global Warming Politics in a Post-Environmental World, www.Breakthrough.org

[7] Meharg, Sara (2009) Measuring What Matters in Peace Operations and Crisis Management, Kingston ON, Canada, McGill-Queens University Press

[8] The Treaty of Versailles was a peace treaty signed on 28 June 1919. As the most important treaty of World War I, it ended the state of war between Germany and the Allied Powers.

resources. Once these resources were depleted, settlers would move on, repeating this cycle. The concept of Manifest Destiny was a persistent source of political division, frequently resulting in conflict over the displacement of Indigenous peoples, the annexation of land by the federal government, and localized wars over water rights. This strategy continued if sufficient resources remained; however, given the finite nature of natural assets, tensions arose as resources diminished and westward migration increased, ultimately leading to localized disputes over land and water rights. Throughout history, global conflicts over natural resources have often been shaped by imperialistic perspectives, a pattern that persists in contemporary discussions concerning environmental and resource-related matters.

Disputes can occur at various levels, ranging from large-scale conflicts such as war, the most extreme form of dispute, to legal disagreements resolved through civil court orders or workplace personnel issues. Environmental disputes occupy an intermediate position within this spectrum. Notably, disputes that escalate into prolonged violence are often more complex and typically possess a lengthy history. The strategies for achieving effective resolution depend on the goals, objectives, and interests of the parties involved. Many environmental conflicts originate from issues like drought, famine, or mismanagement of natural resources, which can intensify tensions to the point of warfare as disputants seek to gain leverage or control.

The imperialist principle, rooted in Roman security strategies, still influences modern policies across cultures and governments. Although such an approach prioritizes preparedness, it may necessitate ongoing measures aimed at addressing potential conflict, which can influence broader efforts toward peace. In countries where military expenditure is substantial, these considerations significantly impact daily life. The issue of balancing peace and defense investment remains pertinent. For example, the global demand for rare earth metals, critical for both consumer products and predominantly for military applications, has become a notable factor within international disputes, including in the ongoing conflict between Ukraine and Russia, where control over these resources is contested.

If we look around the world, the imperialist mindset is represented in most disputes involving natural resources and is demonstrated in many countries' foreign policy, as well as in the way we view our global and national natural resources. Where countries have been locked in bitter battles for hundreds of years, the Roman mindset may be deemed necessary and may evolve into the politics of fear, and fear is an extraordinarily strong tool of persuasion. The role of fear in mediation is both intricate and multifaceted. Although fear can influence how parties position themselves during the mediation process, it must be applied judiciously and alongside other negotiation techniques.

A climate of fear can generate disorder, frequently leading to resource disputes, whereas interest-based systems are designed to address perceived needs. Although alternative dispute resolution literature supports this methodology, shifting such perspectives is inherently difficult, as stakeholders often prioritize self-interest over long-term resource sustainability. The primary challenge for practitioners lies in resolving disputes among parties focused on self-preservation and lacking a collective vision for the future.

To examine natural resource conflicts, which remain a significant cause of war, prolonged disputes, and political instability worldwide, it is necessary to understand history. Given that all nations endeavor to safeguard their assets, it is commonplace for policymakers to advocate for the protection of national interests. The mediator must therefore offer an alternative to prevailing environmental and resource dispute resolution practices by introducing a holistic management process, one capable of fostering transparency, tracking progress, and adapting to evolving conditions. By identifying common pitfalls faced by practitioners in natural resource conflicts, the text elucidates the limitations of existing interest- and rights-based systems in achieving lasting resolution.

The Circularity-Resource-Based (CRB) system proposed herein seeks to stabilize global natural resources by proactively identifying potential conflicts and addressing resource limitations. This integrative approach underscores the connection between environmental challenges and CRB conflict management, thereby facilitating peacebuilding opportunities through international agreements.

The mediator should initiate discussions by emphasizing natural resource conservation as the primary consideration in negotiations related to environmental disputes. I recommend implementing a management framework that prioritizes equitable distribution of resources and sustainable conservation, incorporating the concept of circularity rather than focusing solely on dispute resolution. This strategy leverages existing tools and addresses potential challenges practitioners may encounter during execution.

In most civilized societies, litigation is the standard way to resolve disputes without resorting to war. Courts provide lawful monetary settlements; however, their rulings may not receive universal acceptance. As courts become overloaded and costly, there is a growing shift toward more efficient alternative dispute resolution methods. Many environmental issues are being controlled by the courts using outdated laws. This book aims to guide environmental and resource dispute mediators in developing management strategies that minimize resource impact through recycling and equitable sharing, balancing societal advancement with environmental preservation. Resource disputes are often more complex than typical cases, making standard mediation or co-mediation processes insufficient. Prolonged environmental conflicts require a team-based strategy. This team should include experts in science, engineering, law, and social sciences to tackle the issues comprehensively. Members need strong mediation abilities and must be able to explain technical matters through mathematical models, legal precedents, and regulatory guidelines. This ensures that solutions are balanced and address the underlying causes.

To build this team, we must clarify each professional discipline's role and define key terms such as "success," which varies across fields: engineers and scientists rely on quantitative analysis, while legal professionals often focus on monetary outcomes for clients. It is also important to note that each natural resource necessitates professionals with expertise specific to that resource. In CRB management, success is defined by conserving the resource and maintaining its ongoing benefits, not merely by reaching agreement.

Regulators and attorneys typically communicate using regulatory language and legal logic. Ideally, all disciplines aim to identify the root cause of disputes, though this is rarely achieved through current processes. Success is often measured by a signed agreement or contract, yet such agreements can still be broken. That is why I refer to this approach as alternative dispute management: contracts help manage outcomes, but they do not guarantee them. Practitioners should synthesize these differing perspectives and present information clearly to disputants. In natural resource disputes, true success involves creating enduring solutions that address all parties' concerns and allocate resources fairly for current and future generations. Some mediators consider a signed agreement an indicator of success, although such agreements may at times result in subsequent litigation.

Litigation frequently originates either from the initial parties to a dispute or from additional parties who were excluded from mediation and later adversely affected by the resulting agreement. This indicates that the initial resolution overlooked third and fourth parties. Because natural resources are interconnected, agreements that address only direct disputants may ignore broader environmental impacts and community needs, increasing the risk of overall failure. To gather essential information, the team should conduct a fact-finding mission to identify root causes and all parties involved. In legal proceedings, compulsory discovery tools such as interrogatories, depositions, and document requests serve this purpose, but only after litigation has commenced. While it is preferable for this research to begin before litigation, it often starts later in practice. Attorneys advocate the strongest case for their clients rather than resolving disputes sustainably by conserving resources. Their research typically supports their client's interests, guided by relevant resource laws. As a result, legal resolutions often address immediate concerns but may fail to protect resources for future generations. For example, in western water law, the term "use it or lose it" requires right holders to use their full allotment without regard for conservation, ultimately leading to depletion of the source.

Dispute practitioners should carefully examine disputes to identify underlying causes. If these causes are not addressed, disputes may continue or arise again, affecting relationships over time. Understanding and resolving the true sources of conflict can help prevent repeated issues.

Multiple issues often stem from a series of complex causes shaped by local beliefs and mistrust that may obscure the true root cause(s). These issues are symptoms that may or may not relate directly to the core of the dispute. Treating only the symptoms may offer short-term relief, but the problems persist until practitioners distinguish between symptoms and underlying causes. To address this effectively, practitioners must use methods that uncover both the history and structure of disputes in ways that benefit all parties, society, and the resource itself.

History shows that even seemingly minor disputes can become complicated and escalate over time. Each dispute is unique in its development. Combining legal and scientific expertise offers a comprehensive approach to resolving environmental and natural resource conflicts.

Traditionally, practitioners addressed disputants' needs without considering resource limitations. Identifying resource limitations, however, can prevent future disputes. These skills are essential for both national and international environmental negotiations. Without them, practitioners may overlook underlying resource issues, potentially contributing to even more complex disputes in the future. International disputes do show a growing trend in the conservation of natural resources through the use of Peace Park-type projects[9]. Resolving disputes involving natural resources necessitates both subject-matter expertise in the relevant resource and advanced mediation skills. Mediators must negotiate not only to achieve mutually beneficial outcomes for the parties involved but also to address disparities in natural resource consumption that lie at the center of many disputes. A truly balanced resolution requires meeting current needs while considering the long-term sustainability of the resource for future use. Too often, negotiations are evaluated solely by immediate gains for the disputing parties, which may imply that the resource is either limitless or of lesser value compared to stakeholder interests, an assumption that does not reflect the actual limitations or importance of natural resources.

Given current political, economic, and resource challenges, what might the world look like in fifty years? Will global communities work together to share knowledge and resources peacefully, or will

---

[9] Saleem H. Ali, (2007) Introduction: A Natural Connection between Ecology and Peace, Boston MA. MIT Press.

competition for power and resources fuel ongoing conflict? Although the fall of the Berlin Wall symbolized hope for lasting peace, today's world continues to face multiple, persistent conflicts.

Since the fall of the Berlin Wall, the frequency of global conflicts has risen, often as a result of evolving power dynamics. Competition for limited energy and natural resources has also played a significant role. Many disputes are intensified by climate change, which heightens demand for vital resources such as water. For example, the ongoing crisis in Somalia is linked to prolonged drought periods caused by changing weather patterns. Without a thorough understanding of climate change, inequalities in resource allocation may further fuel international conflicts.

A continued emphasis on priorities shaped by a limited understanding of the interconnections among climate change, social inequalities, and the pursuit of self-interest in natural resources, often framed as economic necessity at the expense of environmental concerns, has the potential to escalate global conflict to levels reminiscent of the nuclear arms race during the Cold War. The ongoing threat of nuclear conflict exhibits similarities to contemporary challenges involving climate change and finite natural resources, as evidenced by scenarios such as Russia–Ukraine and Iran–Israel. These examples highlight the urgency of the issue. It is therefore essential that conflict-resolution professionals leverage their expertise to assist policymakers in resolving disputes concerning natural resources and energy before they intensify. In parallel with conflicts from the Cold War era, present-day U.S. water resource disputes persist, including matters related to western water law, allocations from the Colorado River, fishing rights, mining, and interstate disagreements in the eastern states over Flint River water rights.

Greater attention must be given to natural resources, considering both their importance in the global economy and their limitations. Economic factors play a key role in determining how these resources are distributed. The free-market model is often described as an economic system with limited government involvement, where supply and demand drive market changes. Although this concept is common in introductory economics courses, in reality all markets are subject to some regulation, resulting in economic distortion. These distortions persist because absolutely free markets do not exist anywhere; governments routinely intervene. Despite this, there is widespread promotion of the idea that we are moving toward a free-market system, without fully addressing the consequences, such as resource depletion or the effects of dwindling natural resources on political decisions and global climate change. This shift could result in economic models that foster global conflicts, with growing demand allowing only a select few to access certain resources.

Framing the Issues with Circularity Economy and Environmental Resource Disputes

What is a circular economy?

Environmental circularity focuses on maximizing resource efficiency and minimizing waste throughout production and consumption. It prioritizes reuse, repair, recycling, and designing out waste to keep materials and products circulating for as long as possible, supporting both sustainability and economic efficiency. Currently, we rely on a linear economic model. The linear model is characterized by a "take-make-dispose" approach, whereas the circular economy keeps resources in use for as long as possible while reducing waste. The concept of a circular economy is evident in the management of rare earth metals, where recycling electronics can reduce the need for new mining and processing.

A circular economy aims to keep materials and products in use for as long as possible by minimizing waste and maximizing resource value. We are still early in developing effective recovery and processing systems for rare earth metals; therefore, it is not yet known how many times these metals can be recycled. According to the Save Our Seas 2.0 Act, a circular economy uses system-focused approaches, designs processes to be restorative or regenerative, and prioritizes resource efficiency. Unlike the traditional "take-make-waste" model, a circular economy reduces material use, redesigns products for efficiency, and recaptures waste as resources for new production[10]. A circular economy balances economic growth with environmental protection.

The sustainable materials management (SMM) approach, adopted by the U.S. Environmental Protection Agency (EPA) and other federal agencies since 2009, incorporates principles of circularity. By adopting a circular economy within the SMM framework, efforts are concentrated on lessening materials' lifecycle impacts, cutting hazardous substances, and separating material use from economic growth while meeting societal needs. EPA, in the past, aimed to create a more inclusive and equitable waste management system that addresses climate challenges by introducing strategies for building a circular economy nationwide[11].

This Ellen MacArthur Foundation[12] system diagram illustrates the continuous flow of technical and biological materials through a circular economy.

There are several types of environmental and resource-related disputes, each requiring tailored circularity management plans that can benefit all parties. This text discusses the relevant disputes and

---

[10] https://www.epa.gov/circulareconomy

[11] Basic Information about the Built Environment | US EPA

[12] Ellen MacArthur Foundation's Influence on Circular Economy | Sustainability Magazine

articulates the necessity for designated mediation procedures to ensure outcomes align with conservation objectives and compliance with regulatory standards.

The primary issues summarized below will be examined in greater depth throughout the book. The following concerns are presented from my perspective as they illustrate the connection between ADR and Circularity as a management tool:

1. CRB Management and the ADR Concepts
2. Communication Proficiencies, Approaches, and Processes
3. The CRB Model
4. Professional Challenges, Hurdles, and Pitfalls
5. Mapping of the Nexus
6. Economic Hurdles and Opportunities
7. A Community CRB Model
8. The confluence of science, policy, and the law

CRB Management and ADR Concepts

Alternative Dispute Resolution (ADR) is a concept composed of several distinct types, processes, and forms. Formal and informal mediation each encompass different procedures and expectations, and there is, of course, a gray area in between. The bases on which final decisions are made also differ significantly. Formal ADR process decisions are grounded in law and are binding. Informal ADR decisions may or may not be binding, even if one party asserts that they are. The enforceability of a contractual agreement may depend on which parties are subject to the contract, potentially introducing ambiguity regarding its binding nature. Ultimately, much depends on the circumstances surrounding the mediation.

ADR processes are context-specific in environmental and resource management. Current mediation approaches are not always applicable to environmental and resource disputes because they are often limited in scope and do not address resource conservation. This gap can create a dispute cycle in which parties repeatedly address symptoms rather than resolving the root cause.

Finally, the resource itself may hold key insights into the dispute, and by paying attention to the legal, scientific, social, economic, and political nexus, we can design objective criteria for more successful resolutions.

Communication Proficiency, Approaches, and Processes

Communication is crucial to successful ADR. The way in which the environmental practitioner communicates disputed issues to the parties is paramount. The theoretical mediation approach provides the practitioner with a framework for how the overall mediation will proceed, but the analytical process is the true strength of environmental and resource disputes. Several ADR processes are available; however, the processes selected must provide a foundation for confidence and trust building, making the mediation more productive and successful. By addressing the types of mediation processes, I aim to clarify how and why these systems may or may not work, or why they may offer only limited success, in environmental and natural resource mediation.

As with any specialty, the practitioner must possess advanced skills; not only in mediation approaches but also in the subject matter of the dispute. I have met many mediators who specialize in a particular field because it allows them to focus deeply, anticipate how disputants will position themselves, understand the laws that affect their specific mediation type, and develop systematic processes tailored to that subject matter.

The Resource Model

The complexity of environmental and natural resource issues requires a deeper level of understanding than any other mediation process with which I have worked. Here, I describe how I approach subject-matter analysis and provide case studies demonstrating how court systems are evolving to incorporate different legal doctrines, particularly in water law, to make more informed decisions. The main purpose of presenting a systematic method is to identify the root cause of the dispute through legal and scientific research.

Often, ongoing disputes with superficial issues are mistaken for successful resolution, when in fact the core problem remains unresolved. Unless the practitioner dissects each layer of the dispute and addresses it fully, the conflict will continue and may intensify over time. I compare the practitioner's job to that of a mapmaker: the mapmaker identifies every attribute of the physical landscape and assigns each to a specific layer, then overlays those layers to create a complete geography. Similarly, the practitioner must dissect each symptomatic issue so they can build a model of the dispute that the parties can understand, one that defines the lay of the land.

I approach this process as one might approach a mathematical equation, treating symptoms as variables and considering laws and the physical constraints of resources as constants. Other practitioners may approach it differently based on their expertise, which is why I strongly encourage a multidisciplinary approach. Integrating input from multiple fields enables a clear, comprehensive resolution process that goes beyond contract terms.

Professional Challenges

Mediation practitioners must remember that every advantage carries a corresponding disadvantage. Historically, our governmental system functions best through checks and balances. This method ensures accountability among all parties, maintaining oversight and equilibrium throughout the process. The familiar phrase "trust but verify" captures this principle well. The environmental mediation practitioner must create a process that offers advantages for all parties, including the resource, while building trust through mechanisms that allow for checks and balances, or trust-verified approaches.

This is one of the greatest challenges any mediation practitioner will face, and it becomes even more difficult as the dispute grows in complexity or the number of participants.

Throughout interest-based mediation, practitioners should expect continuous challenges, as disputants consistently pursue their own interests, an inclination that the approach itself inherently encourages.

Regulations, when politicized, can create problematic scenarios in environmental and natural resource disputes. Regulations differ from state to state regarding the allocation of natural resources, whether renewable or not. Certain resource laws originated from common law traditions and did not initially emphasize conservation; resources were often regarded as commodities that could simply be replaced by relocating once depleted. However, resource limitations and consumption patterns have changed significantly since these laws were created.

Chapter 4 provides an overview of some advantages and disadvantages of the current regulatory process. This overview is necessarily limited, as becoming an expert in regulatory matters requires years of legal training and practice. Most environmental mediation practitioners specialize in one or two areas of environmental or resource issues to develop meaningful expertise. It is rare to encounter individuals with comprehensive authority across the full range of environmental topics. Therefore, it is often advisable for the mediation team to include specialists who collectively cover all relevant domains of environmental and resource regulations.

Mapping the Nexus

We often speak of "storytelling" as a method that allows disputants to vent while also documenting their concerns. Storytelling is equally valuable for the practitioner. The way a practitioner organizes and communicates large volumes of information significantly influences the success of the mediation process. In environmental and resource-based disputes, the issues must be distilled into their simplest components, then analyzed for their legal, scientific, social, and political attributes.

The next step is to connect each issue with its attributes in a way that forms a kind of map, one that the disputants can use to establish objective criteria for mutual gain. By thoroughly examining and interrelating each issue, the practitioner enhances transparency throughout the mediation process. This structured clarity helps build trust and confidence among the parties, increasing the likelihood of reaching productive, durable outcomes.

Economic Analysis and Opportunities

The economics of any project are key, critical, and essential to determining its success or failure. An economic analysis can also confer a strategic advantage on one party over another. Because economics plays a substantial role in selecting appropriate dispute-resolution scenarios, it is important to consider how analyses are developed and conducted so that all costs, including those beyond market prices, are evaluated consistently.

Determining the true cost of many environmental and natural resource issues can be challenging, given the complexity of externalities, long-term impacts, and non-market values. The cost-benefit analysis, if not properly applied, can provide for a structural expendability[13]

---

[13] Structural expendability denotes the capacity of a structure to be replaced or repaired without compromising the overall integrity of the system. This concept is commonly applied to primary structural components that are designed for replacement or repair, signifying that substantial maintenance may be undertaken when necessary.

of resources and communities[14]. The structural element of economic evaluation emerges from how the cost-benefit analysis is constructed. For example, if a cost-benefit analysis relies solely on the market value of a resource as a positive cost while ignoring negative costs, such as cleanup, mitigation, and decreased property values, the analysis will likely justify the mining or use of that resource without acknowledging its limitations or the broader consequences of extraction. The way an economic analysis is framed can therefore give individual disputing parties unfair or biased leverage over others. For this reason, it is essential that the practitioner ensure an equitable environment in which costs are assessed objectively and strategies remain balanced.

The Confluence of Science, Policy, and the Law

Environmental and resource mediation is one of the rare fields in which the full spectrum of scientific disciplines, legal frameworks, and political ideologies converge for the purpose of resolving disputes. Successfully addressing conflicts involving environmental and natural resources often requires drawing on all relevant areas of expertise, depending on the complexity of the matter. These diverse knowledge streams, science, law, and policy, merge like tributaries into the large river of mediation.

The objective of the CRB process is to integrate sound science, effective policy, and applicable law into the mediation framework. By bringing these disciplines together, the CRB process enables disputants to design their own resolutions based on reliable information and grounded principles. This approach enhances mediation outcomes while supporting resource conservation through circularity concepts.

---

[14] Steady, Filomina, Environmental Justice in the New Millennium, New York, Macmillion, (2009)

Chapter 2
CRB Management

As more conflicts over natural resource scarcity become known, the potential for these conflicts to negatively affect global and local economies increases, while also creating opportunities for violence on the world stage. As long as global interests continue to conflict, the need for resolution becomes increasingly urgent. Only when we treat these issues as a priority will they receive the consideration they truly deserve. Given the rising cost of litigation, many practitioners have turned to alternative processes that have produced more positive outcomes for all parties involved. These global conditions create a unique opportunity for environmental and resource conservation and preservation through a systematic CRB mediation management system, provided it is applied by practitioners with the appropriate expertise.

The court system offers a formal process to resolve disputes between parties, often resulting in a definitive outcome. In certain cases, however, both parties may experience resource depletion as a result. Within formal court proceedings, decisions are made according to the rule of law or by judges applying existing legal principles, which may not always account for contemporary concerns such as environmental sustainability. Court decisions are also typically final, leaving little room for adaptive or long-term solutions.

Currently, two primary negotiation methods are used to handle environmental and natural resource disputes: interest-based and rights-based approaches. Professionals working in this field must have a solid understanding of these strategies and apply their conflict-prevention skills to help parties identify problems early and resolve resource and energy disputes before they escalate. Greater emphasis should be placed on the resources themselves, along with a deeper understanding of their role in economic and political stability. When disputes are managed through systematic approaches, they can become opportunities for more effective and enduring solutions. Moreover, adopting a third system, such as the CRB management model, may better address the unique challenges of natural resources and promote sustainable, long-term outcomes.

Natural resource disputes are particularly difficult to resolve because they often involve multiple parties, complex legal issues, and secondary factors that divert attention from the root cause of the conflict. These challenges are further intensified by increased public attention, which introduces additional concerns, and by differing views among participants regarding what constitutes a "successful solution." Practitioners can turn public involvement into an advantage by fostering transparency, building trust, and supporting the dispute resolution process. Success in natural resource disputes should be measured by ensuring that resources remain available for future generations, as defined by the Public Trust doctrine. Therefore, in cases such as freshwater disputes, success should not be limited to reaching an agreement among current users. Instead, it should focus on achieving an equitable outcome that fairly distributes liabilities and benefits among all stakeholders, including consideration for the resource itself.

There are countless ways to mediate individualized conflicts. No single mediation method can resolve every dispute, as each situation presents unique facts, issues, parties, resources, and legal considerations. As a result, mediators must employ a structured process supported by analytical tools that account for the costs associated with proposed resolutions. While solutions may take various forms, the most effective ones are those that integrate economic, environmental, and social costs. In cases like water disputes over natural resources, creative mediation strategies can lead to better outcomes when traditional approaches have failed. [15]

As global populations continue to grow, natural resource consumption increases, leading to more frequent disputes over these limited and essential assets. To conserve resources and maintain their quality, efficient and effective problem-solving methods are necessary. Past examples show that current mediation processes often fail to resolve resource-related disputes because they do not adequately consider the availability or limitations of the resource itself, factors that should serve as key guideposts for resolution. Without a full understanding of resource availability and its impact on the parties' ultimate settlement, the likely outcomes include unsuccessful implementation, overuse or misuse of the resource, repeated mediation, litigation, wasted time and money, and dissatisfied parties. By continuing to rely on these approaches, we move further in the wrong direction. Rather than seeking resolution through traditional methods and litigation, an innovative approach is required to fully resolve natural resource disputes, particularly water-related conflicts. The proposed resource-based mediation process facilitates solutions by encouraging a comprehensive understanding of the resources and the interests of involved parties, which leads to the implementation of long-term agreements and the management of the resource. [16]

To prevent disorder during mediation, I observe that disputes typically revolve around three fundamental elements: interests, power,

---

[15] Ruhl, J.B., (2003), Equitable Apportionment of Ecosystem Services: New Water Law for a New Water Age, Vol. 19:1, Journal of Land Use

[16] Ruhl, J.B., (2005) Water Wars, Eastern Style: Divvying Up the Apalachicola-Chattahoochee-Flint River Basin. *Journal of Contemporary Water Research & Education, No. 131, 2005*
FSU College of Law, Public Law Research

and rights. These elements form the foundation of most current mediation systems. However, such approaches often overlook resource limitations and instead prioritize market value. While many systems advise mediators to focus on one or more of these elements, I argue that, to maintain order and effectiveness in the process, mediators must address all of them, explicitly incorporating resource constraints, and establish a structured method for evaluating each foundational element.

The interest-based system was formulated by Ury, Brett, and Goldberg[17] and places the focus on the individual interest[18]. It is also important to note that the reason stated by these authors for coming up with such a system was that the systems used at that time were consistently inefficient[19]. How can we identify when a mediation process is ineffective? My approach begins by using success as the primary measure of a mediation's effectiveness. This raises an important question: how do we determine whether a dispute has truly been resolved, rather than merely satisfying individual interests without accounting for limited resources? If those resources are exhausted in the process, it is necessary to ask whether the outcome can still be considered a successful resolution.

An effective dispute resolution system should function efficiently, address relevant interests, sustain long-term stability, and prevent deterioration into disorder or unmanageability. Equally important is the clear definition and demonstration of criteria for genuine success, along with methods for measuring and monitoring outcomes over

---

[17] Ury, W., Bret, J., Goldberg, S., getting Disputes Resolved: Designing Systems to Cut the Costs of Conflict, (Jossey-Bass, London (1988)

[18] Id. Roger, F. and Ury, W., Getting to Yes, Negotiating Agreements without Giving In:1997

[19] Id. Ury, W., Bret, J., Goldberg, S., (1998)

time. Although confidentiality agreements often limit the collection of national mediation success data, practitioners can still evaluate their own effectiveness. To support this evaluation, current resolution systems must be critically examined.

Defining Success

The purpose of defining success is to equip environmental practitioners with a structured methodology designed for a mediation team approach to resolving environmental and natural resource disputes. This process should be methodically organized into a sequence of incremental stages, each building on the previous one to create an integrated and resilient framework. Progression from one stage to the next requires consensus among all involved parties, with each stage subject to thorough evaluation.

Embedded within this framework are conflict prevention techniques intended to identify and address potential points of contention or symptomatic issues early in the process, thereby reducing escalation related to underlying or third-party concerns during mediation. Throughout the process, the objective is to collaboratively determine what participants believe to be the fundamental cause of the dispute, while acknowledging that perceived root causes may, in fact, be symptoms. Building an effective process team also requires careful consideration of the unique contributions and communication styles of various professional disciplines. Additionally, “success” must be defined in terms relevant to the specific natural resource or environmental issue at hand, recognizing that true success extends beyond simply obtaining signatures on an agreement.

Different scientific and legal disciplines define success in distinct ways and at varying levels. Engineers may rely on mathematical or statistical tools, while others emphasize logical reasoning or legal frameworks. Political professionals do not always follow the same conventions. Within science, some fields prioritize quantitative expression, whereas others depend on observational evidence. Regulators and lawyers typically ground their analyses in regulations that support structured legal arguments. To facilitate progress, practitioners should encourage open dialogue among all participants, allowing diverse perspectives to be integrated into a balanced oversight process. Ultimately, all disciplines seek to identify the root cause of disputes. In the context of natural resource conflicts, true success lies in achieving a durable resolution that addresses shared concerns and allocates resources sustainably for both present and future generations.

Environmental and natural resource professionals must increasingly possess advanced negotiation skills that integrate policy, scientific, and economic expertise, along with strong issue-spotting capabilities across various ADR (Alternative Dispute Resolution) approaches. An organized issue-spotting methodology, one that incorporates policy considerations, economic management, and negotiation strategies beyond traditional interest- or rights-based systems, is essential for producing well-balanced technical solutions with lasting benefits. Practitioners should review identified issues collaboratively with all parties to ensure their relevance and completeness. Historically, resolution efforts have focused on solutions tailored to disputants' immediate needs, often without regard to resource constraints or root causes. However, environmental and natural resource disputes frequently involve interests broader than those of any single party. Practitioners therefore have a responsibility to address these concerns in ways that engage all relevant stakeholders throughout the process.

Natural resource disputes can also have global impacts. Resolution of international disputes in Africa illustrates a growing trend in the conservation of natural resources though the use of Peace Park types of projects[20]. Resolving disputes involving natural resources requires specialized expertise and carefully designed mediation processes. In cross-border resource conflicts, solutions should aim for mutually beneficial outcomes that balance resource consumption with strategies supporting local economic development. Effective resolution must consider not only the immediate needs of the disputants, but also the long-term sustainability of the resource itself. Too often, negotiations are evaluated solely on the gains secured by the parties, implicitly assuming unlimited resources or undervaluing their significance relative to short-term interests, assumptions that may be flawed.

Where applicable, a CRB system can contribute to the stabilization of global natural resources by identifying traditional points of contention early in the process and emphasizing resource limitations. In doing so, it reveals a natural and legal nexus between environmental concerns and conflict resolution mechanisms. Natural resource nexus provides opportunities for peace building, through the construction of global agreements[21].

Most mediation practitioners structure the process to focus on key points of dispute and establish pathways toward resolution. The chosen approach may vary depending on the subject matter and whether the dispute occurs in a formal or informal setting. These contextual factors also influence whether the mediator assumes a more passive or active role in guiding the process.

---

[20] Id. Saleem (2007)

[21] Id. Saleem (2007)

Fig. 2.1

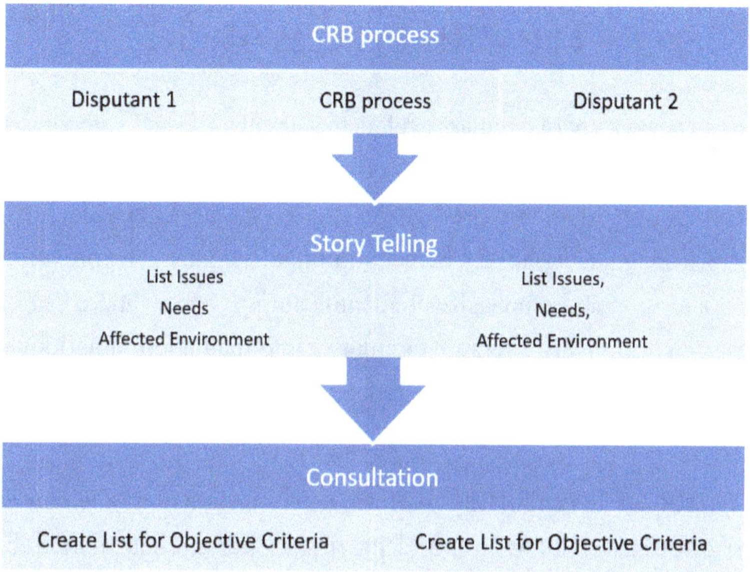
CRB process
Disputant 1
CRB process
Disputant 2
Story Telling
List Issues
Needs
Affected Environment
List Issues,
Needs,
Affected Environment
Consultation
Create List for Objective Criteria
Create List for Objective Criteria

Chapter 3
The CRB Process Model

The concept of circularity refers to a circular lifecycle of industrial practices that maximize resource utilization and minimize waste throughout the entire production and consumption cycle, while prioritizing sustainability and economic efficiency. The prevailing circularity model emphasizes remanufacturing, repair, and beneficial use. To effectively integrate circularity into mediation, practitioners must understand and apply circularity principles as incentives for disputing parties.

The UNEP circularity platform offers insights into the concept of circularity, its scope, and its role in advancing sustainable consumption and production patterns[22]. UNEP's focus on circularity and sustainable consumption and production is central to achieving multilateral agreements ranging from the Sustainable Development Goals and the Paris Agreement to the post-2020 global biodiversity framework. These principles are also essential for fostering a sustainable recovery in the aftermath of the COVID-19 pandemic. In addition, natural disasters can act as catalysts for addressing social challenges by encouraging more effective natural resource management, ultimately contributing to improved local economies. Circularity relies on value-retention loops, as illustrated by the UNEP circularity approach below:

[22] Circularity | UNEP - UN Environment Programme

Fig. 3.1

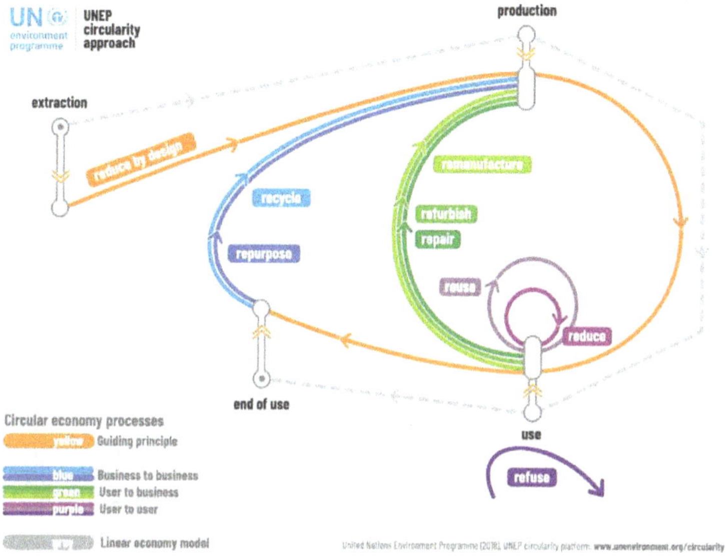

This chapter presents examples illustrating how regulations and beneficial uses may be applied to natural resources and environmental issues, such as water rights and coal waste, and how these factors can be evaluated by disputants to achieve greater success. To begin, it is necessary to review various ADR models so practitioners can determine which approach best fits a particular dispute. In practice, a practitioner may need to draw on elements of both interest-based and rights-based ADR systems to develop a more effective resolution.

As discussed earlier, multiple mediation processes are available, but determining which is most effective requires careful assessment. Practitioners must match the appropriate process model to the specific dispute at hand. The goal is to distinguish underlying causes from the visible issues and symptoms of a conflict and then select a dispute resolution approach accordingly. Making the correct choice significantly increases the likelihood of meaningful success. For example, in disputes involving straightforward contractual or social issues where legal rights predominate, the CRB model may be less effective than an interest-based approach. Rights-based and interest-based models are generally most effective when a limited number of parties are involved and when a mutually acceptable agreement is attainable. In contrast, environmental and natural resource disputes present distinct challenges and require a more comprehensive examination of both the dispute and its broader impacts, including considerations for the environment and future generations. The CRB model is particularly valuable when environmental remediation and material recycling are possible, as these factors provide additional incentives for all stakeholders.

With continual increase in global population that is directly affected by an increased demand for natural resource consumption, natural resource disputes are inevitable.[23] Natural resource disputes are inherently complex and difficult to resolve. They often expand to include peripheral issues, additional parties, and increased public scrutiny, where the definition of a "successful solution" may vary widely. At a minimum, success in resolving natural resource disputes should be measured by the retention and protection of the resource itself. Because natural resources are limited yet essential for survival, more effective problem-solving approaches are needed to preserve both their quality and quantity. The CRB model was developed to address the shortcomings of prior resource dispute mechanisms, facilitate sustainable and enduring solutions, and ensure both resource conservation and equitable economic incentives for all stakeholders. While the model is broadly applicable to many natural resources, this chapter focuses primarily on water and coal.

---

[23] Meharg, Sara, Measuring What Matters in Peace Operations and Crisis Management, (2009).

By defining a resource's limitations through an examination of its lifecycle and generating resolution options based on that standard, practitioners can more effectively protect the mutual interests of disputants and other stakeholders. Exploring new and innovative processes, while retaining functional elements of prior models, allows for a framework that more accurately evaluates the true costs associated with consuming limited natural resources and managing renewable ones. Unlike earlier, less effective approaches, the definition of "success" under the CRB model is both objective and subjective: it not only conserves the natural resource but also delivers practical solutions that satisfy the parties' genuine needs.

Circularity is often undervalued as a tool for resolving natural resource disputes. However, it offers the potential to generate a broader range of solutions than litigation, which is frequently constrained by outdated legal doctrines. The use of alternative approaches such as circularity can yield positive and promising outcomes for those who choose to adopt them. Benefits include trust and the ability to provide the community with a sense of justice.[24] Unfortunately, basic circularity approaches do not address all factors surrounding resource and water disputes. Failing to consider these factors hinders the ability to establish a successful resolution.

THE CRB MODEL

The CRB Model serves as one of several tools specifically designed to systematically analyze resource disputes and present disputants with scenarios grounded in objective criteria. Within this framework, all technical and legal matters are openly and transparently discussed, enabling the parties involved to select from a range of alternatives

---

[24] Id. Meharg, supra.

for resolution. In practice, this model provides comprehensive guidelines for conducting the holistic analyses essential to resolving resource-related conflicts.

Procedurally, the CRB Model is most effective when implemented through a facilitated mediation approach. In this process, disputants evaluate objective criteria under the guidance of a trained practitioner team. Among Alternative Dispute Resolution (ADR) methods, facilitated mediation stands out for its flexibility, allowing parties to pursue a tailored process or outcome aligned with their unique needs, genuine interests, and the specific constraints of the resource in question.

Building on the four foundational principles of the interest-based model, the CRB Model further incorporates rights-based perspectives, along with social, legal, technical, environmental, and socioeconomic considerations, into the formulation of objective criteria used to develop resolution scenarios. These criteria are developed by practitioners to inform disputants of the core issues necessary for achieving durable resolutions.

As an alternative approach, the CRB Model equips parties with essential tools to address complex disputes and facilitates more informed decision-making. By establishing an objective criterion process that reflects all critical factors relevant to the resource, the CRB Model serves as a strategic guide toward viable solutions. Thoughtful analysis of these factors during the creation of objective criteria enables disputants to better understand underlying challenges and anticipate potential future concerns.

Information pertinent to the dispute is collected by a practitioner or mediation team serving as neutral parties. Accordingly, practitioners must possess a thorough understanding of the mediation process and relevant expertise regarding the resource and its associated legal considerations. The ability to formulate effective agreements depends both on the practitioner's competencies and the inherent limitations of the resource involved. To facilitate resolution, practitioners prepare thoroughly, clarify constraints affecting the resource, assess how these limitations influence the specific dispute, and develop solution options that align with established guidelines.

Environmental practitioners must also demonstrate adaptability in response to dynamic circumstances. It is recommended that assessments begin from a balanced perspective, thoroughly considering both the limitations of the resource and the influence of internal and external factors on dispute progression or resolution. The complexity of this approach increases when dealing with multiple resource types that involve numerous variables. To ensure an equitable framework, all inputs and outputs should be incorporated into the model. Additionally, a structured formula should be developed to account for forces or issues impacting the equation, including:

1. Actual economic costs, including both benefits and detriments
2. Legal and regulatory obligations
3. Constraints and availability of resources
4. Public and community input
5. Resource circularity

Data gathered through issue identification and collaborative processes should be systematically collected by the practitioner or their team. This methodical approach supports disputants in establishing objective criteria for informed decision-making.

Fig 3.2

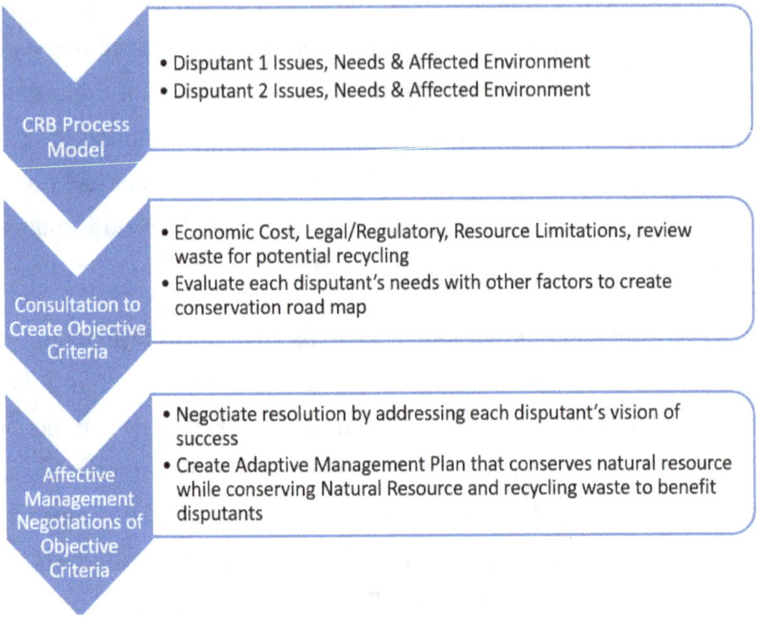

Establishing Resource Objective Criteria Using CRB

The practitioner must methodically analyze the dispute as if it were a complex mathematical equation. In this framework, the variables represent individual issues, while the constants reflect constraints established by legal doctrine and the availability of natural resources. By demonstrating how changes to these variables and constants alter overall outcomes, the practitioner prepares scenarios that help disputants understand and develop objective criteria. These criteria are then presented during mediation and used as a foundation for generating resolution options. The primary objective at this stage is to establish an unbiased standard that all resolution options generated by the parties must satisfy.

The practitioner develops the objective criteria evaluative process based on the following constants:

1. Resource limitation
2. Laws and regulations
3. Economic factors
4. Scientific and engineering considerations
5. Parties' actual needs
6. Other factors relevant to the specific resource

Using these factors as a threshold ensures that future resource use allows for conservation or recharge while remaining within applicable legal boundaries. The practitioner then enhances the evaluative process by introducing variables related to resource limitation, including:

a) Physical, chemical, and biological characteristics
b) Temporal effects
c) Regulatory effects (e.g., water quality) of changing regulations
d) Direct and indirect impacts on upstream and downstream users
e) Applicable laws, including state, interstate, and federal laws
f) Changing legal doctrine
g) Unintentional impediments created by doctrine
h) Judicial opinion

A thorough understanding of state, federal, and international environmental regulations enables practitioners to establish objective criteria that benefit all parties while supporting environmental stewardship. In some mining cases, process waste can generate additional revenue through recycling initiatives, such as recovering Critical Rare Earth Metals (CREM) from Coal Combustion Residuals (CCR) and hard rock mine tailings. These examples illustrate how leveraging regulatory programs can create valuable opportunities.

On October 12, 1980, Congress enacted the Solid Waste Disposal Act Amendments of 1980 (Public Law 96-482) to the Resource Conservation and Recovery Act (RCRA), incorporating the Bentsen and Bevill Amendments (Sections 3001(b)(2)(A) and 3001(b)(3)(A), respectively). The Bevill Amendment provides exemptions for: (1) fossil fuel combustion waste; (2) waste resulting from the extraction, beneficiation, and processing of ores and minerals, including phosphate rock and overburden from uranium ore mining; and (3) cement kiln dust. Consequently, beginning in 1986, the EPA deferred regulation of coal mining and coal beneficiation wastes to the Department of the Interior under the Surface Mining Control and Reclamation Act (SMCRA). Waste generated at coal mines, including coal beneficiation wastes and overburden, is therefore regulated pursuant to SMCRA.

Recent approaches under the RCRA Bevill Amendment regarding CREM recovery indicate that raw coal is not classified as RCRA waste because it is a recognized valuable commercial product. Off-specification coal rejected by preparation plants often retains commercial value, although it may be considered waste if it is disposed of. Similarly, solids that settle out in wastewater treatment plants associated with coal preparation frequently possess commercial value and may be blended with product coal; these solids are typically not designated as RCRA waste unless they are disposed of. Department of Energy research suggests that CREM recovery from raw coal and off-specification coal may involve processes such as crushing, sizing, and leaching. The recovery of CREM from coal, off-specification coal, and preparation plant solids falls within the scope of the Bevill Amendment, which extends to all ores and minerals. Furthermore, Bevill rulemakings clarify that beneficiation includes leaching; thus, waste generated from leaching these materials may be exempt from regulation under RCRA Subtitle C, provided the leaching process meets the regulatory definition of dissolution (see 54 FR 36618), rather than digestion.

Coal by-products, collectively referred to as coal combustion residuals (CCRs), include fly ash, flue-gas desulfurization (FGD) solids, bottom ash, and boiler slag. On December 19, 2014, the EPA Administrator signed the CCR Rule into effect, establishing technical requirements for CCR landfills and surface impoundments under RCRA Subtitle D. This rule permits the continued beneficial use of CCRs in construction and other applications.

In 2015, the EPA issued the CCR Rulemaking (80 Federal Register 213012), establishing criteria for the beneficial use of CCR (65 FR 32230). As long as the criteria outlined in 40 CFR § 257.53 are met, future beneficial uses of CCR are not limited by the rule. Recovering CREMs from CCR appears to meet these beneficial-use standards. When CCR is removed from impoundments or landfills and processed, typically through leaching and precipitation, to create a CREM concentrate, the activity is considered a beneficial use due to existing domestic and foreign markets for rare earth element (REE) concentrates. However, if coal ash used for CREM extraction is removed and the landfill or impoundment is later found to cause environmental contamination, the responsible company may face liability under Section 7003 of RCRA's strict-liability provisions.

Traditionally, electric utilities have relied on coal as their primary fuel source for electricity generation. This reliance has produced thousands of tons of CCR, requiring costly disposal in landfills without addressing the underlying issue. Research conducted by the DOE has shown that CCR contains CREMs at concentrations feasible for economic recovery. By processing CCR to extract CREMs, utilities can reduce landfill volumes, provide CREMs as a usable resource, potentially generate financial benefits that lower operating costs, create local jobs, and reduce environmental risks associated with accidental releases.

The Good Samaritan and the Comprehensive Environmental Response, Compensation, and Liability Act (CERCLA)

The Good Samaritan Remediation of Abandoned Hardrock Mines Program (Good Sam) is a federal initiative designed to facilitate the cleanup of abandoned hard-rock mine sites. The Act, which became law in December 2024, allows Good Samaritans, typically mining companies, to apply for permits to investigate or remediate environmental hazards at qualified mine sites without incurring certain liabilities under CERCLA or the Clean Water Act. Permit holders are granted exemptions from select requirements, provided they comply with the terms of the permit. The Act authorizes up to fifteen remediation permits and fifteen investigative sampling permits. The EPA manages permits for projects on private, state, or Tribal lands, while projects on federal lands are reviewed by the appropriate federal agency.

For abandoned coal mining sites, CERCLA liability is further complicated by the Surface Mining Control and Reclamation Act (SMCRA), which often does not apply to sites abandoned before 1977 (pre-law sites). Redevelopment or reuse efforts at these locations can trigger liability if contaminants are released. According to the EPA, a Good Samaritan is any group or company willing to clean up orphaned mine sites that it does not own. Existing guidelines clarify that Good Samaritans may avoid CERCLA liability when recovering resources such as rare earth elements (REEs); however, they may still be subject to Clean Water Act liability if pollution continues after cleanup activities are completed.

Hard-rock mine tailings, like coal and CCR facilities, contain usable levels of CREMs. Many abandoned mines in western regions pose ongoing risks of surface- and groundwater contamination. The

Good Sam program functions as a demonstration project focused on best practices for recovering CREMs from these sites. Extracting CREMs from abandoned mines offers benefits similar to those associated with coal and CCR recovery and may also reduce the need for new mining activities.

Economics

The criteria needed to evaluate variable economic factors also vastly differ from project to project and can be difficult to evaluate. There are many tools that provide market value for natural resources, but few that demonstrate the actual cost. Many economic tools address the limitations of a resource by increasing pricing, while other economic tools present a bias based on the concept of "Free Market" externalities. Externalities are an economic concept where a cost is associated with a good or product.[25] In the case of a natural resource, outside market forces are generally not reflected in the price of the resource, which creates an economic distortion for which the practitioner must be aware. "Economic distortions arise because there really are no free markets operating anywhere in the world, since governments of all types frequently intervene in the marketplace."[26] Additionally, economic models can be constructed in such a way as to define a resource or a community as "structurally expendable,"[27] as is the case with environmental justice disputes, where the resource or a community is expendable based on its economic model.[28] "*Emergy*"[29] is an

---

[25] Id. Stewart (2009)

[26] Id. Stewart (2009): 176-178

[27] Id. Steady (2009)

[28] Id. Steady (2009)

[29] Id. Odum (2007

example of an ecological/economic concept brought forward by H.T. Odum, a world-class ecologist, which combines ecology and economics to demonstrate the true value of ecological models. Emergy applies a systematic approach to incorporate the concepts associated with thermodynamics with the principles of ecology and economics to achieve the actual cost of that resource in an attempt to remove the issues of economic distortion.

International disputes illustrate a growing trend in the conservation of natural resources through the use of Peace Park[30] type projects and the concept of *Emergy,*[31] which provides for an alternative approach to determine the value of natural resources. There are several examples where "environmental conservation zones can facilitate the resolution of territorial conflicts."[32] The nexus is both economic and environmental, in that a disputant may profit monetarily from conservation of the natural resource, while the resource itself benefits from the conservation efforts of the parties.

Market forces direct and indirect externalities

a) *Emergy.* Renewable natural resources function within natural cycles, such as forestry. A forestry resource has a logical starting point and end that correspond to its life cycle. Howard T. Odum authored numerous influential works on ecology and is widely regarded as a world-class ecologist. Odum introduced the concept of "emergy" and ecological cycles, which demonstrate the balance required for ecological and energy cycles to reach completion. By applying Odum's ecological concepts to the field of

[30] Id. Saleem, Ali, (2007)

[31] Id. Odum (2007)

[32] Id. Saleem (2007)

dispute resolution, it is possible to demonstrate the nexus that exists in nature as well as in environmental and resource dispute resolution. There exists a balance in everything that we do as demonstrate by Odum's many ecological texts reference[33]. When Odum's concepts are applied to negotiation, the negotiations themselves can be understood as a type of cyclical model dependent upon balance within the process. The professional environmental mediation practitioner must therefore keep the concept of balance in mind when negotiating disputes. The goal is to integrate dynamic natural resources into the CRB process so that solutions emerge through a feedback model. This feedback model supports settlements based on objective criteria that benefit the disputants while avoiding disruption of the resource's delicate natural balance. In this way, balance encompasses the interaction of energy and market forces with ecological cycles, as well as balance within the negotiation process itself. The natural nexus exists only when all forces, economic and environmental, are engaged in preserving natural resources for future generations and when alternative sources are given equal weight in the negotiation process.

The overall goal is to provide objective criteria based on true economic value, supported by dependable scientific and engineering principles and coupled with applicable law, thereby creating an all-encompassing foundation for resolution.

Mediation Preparation

The preparation stage allows the practitioner to gather as much information as possible to ensure a smooth mediation process.

---

[33] Id. Odum et. al

Practitioners should conduct early individual meetings with disputants to ask questions and actively listen to explanations and responses surrounding the dispute. Practitioners with strong issue-spotting, active listening, and questioning skills tend to be more effective throughout mediation and are more likely to help disputants reach a successful resolution.

Individual meetings are essential to understanding each disputant's concerns, behaviors, positions, interests, and goals. As an added benefit, the CRB Model permits practitioners to collect all relevant information for use in mediation without evidentiary restrictions. Practitioners may use a "Conflict Triangle" when necessary to review who each disputant is, how disputants interact, and identify each disputants' alleged cause of the dispute.[34] This is a valuable tool during preparation to analyze future productivity and determine whether mediation is appropriate for this dispute. Evaluation of the disputants' willingness to negotiate provides insight into how the practitioner might foster mutual resolution.[35]

Evaluation of the disputant's willingness to negotiate provides opportunities for how the practitioner might foster mutual resolution.[36] Early research into the dispute history is also crucial to ensure that proposed solutions do not repeat past mistakes. Identifying the parties' previous failed attempts at resolving similar disputes, understanding what contributed to those failures, and avoiding those pitfalls during

---

[34] BEER, J., et al., (1997) The Mediator's Handbook (3d ed).

[35] COLEMAN, P.T. et al., (2008) Reconstructing Ripeness II: Models and Methods for Fostering Constructive Stakeholder Engagement Across Protracted Divides, 26 Conflict Resol. Q. 1.

[36] Coleman, P. T, Fisher-Yoshida, B., Stover, M.A., Hacking, A.G. Bartoli, A, Reconstructing Ripeness II: Methods for Fostering Constructive Stakeholder Engagement Across Protracted Divides, Conflict Resolution Quarterly, Vol.26, No. 1, Wiely Periodicals, INC. (2008)

mediation allows disputants to develop alternative, and potentially more successful, solutions.

To fully understand the natural resource issues involved, the practitioner must apply a balanced approach. This approach may include, but is not limited to, the following evaluations:

1. The resource's physical, chemical, and biological limitations.
2. A systematic assessment using sound engineering, science, and law to distinguish symptoms from underlying causes.
3. Identification of whether external issues create symptoms without losing focus on the root cause.
4. Examination of how outside forces affect potential solutions.
5. Development of balanced solutions based on objective criteria that promote mutual gain.

As part of this process, the practitioner must uncover individual interests, existing legal rights, and desired outcomes without allowing these elements to become the primary focus of the mediation.

Practitioners may then establish mediation procedures with input from the disputants. These procedures can be tailored to meet the specific needs of the parties. After researching and collecting the necessary information, the practitioner will explain the mediation process moving forward.

Establishing the CRB Model

Once sufficient information regarding the disputants' interests, resource conservation concerns, and mediation goals has been collected, the practitioner may begin considering additional factors related to the water body or water source at issue that could affect the long-term success of any implemented agreement. An objective criterion is a mutually acceptable standard derived from credible, unbiased sources. Use of objective criteria during negotiations is quite useful because "[d]decisions based on reasonable standards make it easier for the parties to agree and preserve their good relationship."[37] Although it may appear that objective criteria anchor one all-encompassing solution, it actually provides more of a playing board for disputants to discuss and develop multiple options to resolve the dispute. Development of the objective criterion in the CRB Model addresses four major categories relating to the resource. Considerations include the resource's physical properties, applicable laws, economic factors, and disputant interests.

See Fig. 3-3

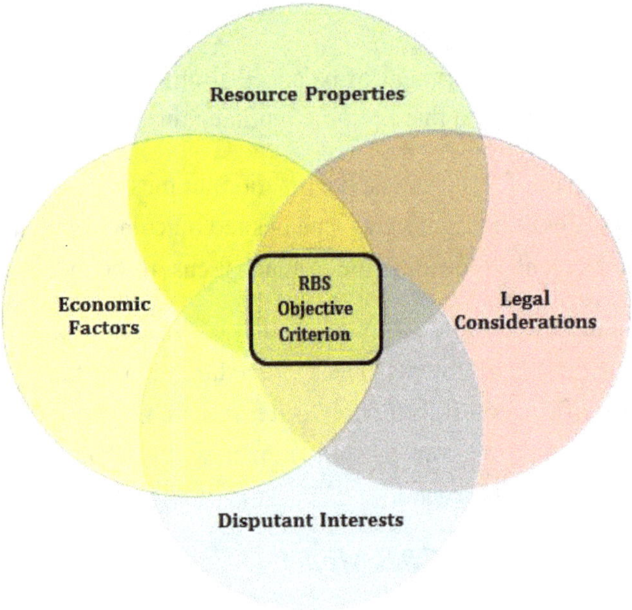

[37] ROGER FISHER & WILLIAM URY, *Getting to Yes: Negotiating Agreements Without Giving In* (2nd ed. 1983).

In effect, the use and formation of the objective criterion within the CRB Model separates this mediation model from other problem-solving approaches. The model permits a broader analysis rather than primarily focusing on singular considerations, such as individual interests. Focusing solely on the disputant's needs leads only to partial resolution, and a dispute cannot be fully resolved if the resource no longer exists.[38] The use of these factors as a threshold ensures that future conservation of the resource will remain within the legal boundaries, if any.

Resource Properties

The characteristics of resources must be evaluated to maintain both availability and quality. In the context of water disputes, a watershed model approach can be used to illustrate current usage and recharge rates of a water body, as well as its chemical and biological attributes. Watershed models incorporate factors such as precipitation, land-use type, impervious areas, slope, soil types, and drainage area. This approach allows potential outcomes related to increased use and changes in water quality to be anticipated, which may help prevent excessive consumption or deterioration of the resource. Depending on existing allocations of the water body, both direct and cumulative effects should also be assessed. A practitioner or external expert may develop a watershed model for the specific water body under consideration.

Governmental regulatory agencies, for example, utilize similar models to conduct water quality studies and perform watershed management.[39] A Geographic Information System (GIS) is a valuable

---

[38] J.B. RUHL, *Water Wars, Eastern Style: Divvying Up the Apalachicola-Chattahoochee-Flint River Basin*, 131 J. Contemp. Water Res. & Educ. (2005).
[39] U.S. ENVTL. PROT. AGENCY, *BASINS (Better Assessment Science Integrating point & Non-point Sources)*, (visited December 10, 2013), *at* http://water.epa.gov/scitech/datait/models/basins/. [12] *Id.*

tool used by regulatory agencies to provide information pertaining to a given area. GIS programs organize spatial and temporal data so it can be displayed as maps, tables, and graphics. These programs allow practitioners or their consultants to analyze landscape information and illustrate relationships among data sets. Water body data obtained from regulatory agencies can be added as layers to the watershed model, enabling practitioners to assess resource availability and quality for stakeholders. In many cases, the primary purpose of using GIS is to visually represent the resource and provide a comprehensive overview.

The CRB Model offers a systematic analysis specifically tailored to resource disputes. In practice, dispute analysis under this model incorporates a holistic view of the conflict, rather than creating narrow focal points, as is often the case in traditional ADR and court-based models. Through this analysis, key "guideposts" effectively establish objective criteria and form a baseline for disputants. Settlements developed using this baseline are more likely to result in successful long-term solutions, as disputants account for factors that may affect their resolution in the future.

This process is achieved through thorough preparation, identification of resource limitations, and assessment of how these constraints may influence the resolution options under consideration. Practitioners then develop solution options that align with established guidelines. In addition, Artificial Intelligence (AI) may be used to assess conditions and generate a range of potential solutions for disputants to evaluate.

Legal Considerations

The objective criteria established under the CRB Model must operate within the boundaries of the law. Applicable federal, state, and local laws must be considered to ensure that future agreements comply with legally established resource uses. When disputes extend across multiple states, relevant laws or doctrines governing the resource may conflict, and the practitioner must address these discrepancies before an agreement is finalized.

Practitioners must also consider the legal rights of disputants and other users. In some cases, the law permits or even encourages additional resource consumption. Practitioners are therefore advised to move beyond a strict "first in time, first in right" approach while still recognizing established water rights. When allocation issues arise, methods similar to those used by Indigenous communities may offer useful guidance. Research shows that Indigenous populations have addressed water scarcity through strategies such as time-based allocation, prioritized use, and protection of downstream access. For example, Berber communities in Morocco's High Atlas Mountains have resolved water conflicts by allocating use based on time rather than volume, thereby distributing risk more evenly among users. These communities also distinguish among different types of use to reduce waste. Although prioritizing use can be complex, international treaties incorporating such principles do exist, and these mechanisms may be applicable to contemporary disputes.

Equitable distribution of downstream rights is an approach the United States is still learning. After several years of unsuccessful negotiations and court cases, water conflicts between the State of Florida and Georgia over the Apalachicola, Chattahoochee, and Flint ("ACF") River Basin remain unsettled.[40] In fact, the State of Florida recently filed a complaint in the United States Supreme Court seeking equitable distribution and injunctive relief, alleging that the State of Georgia is excessively consuming water from the ACF, thereby

---

[40] *See* RUHL, *supra*.

detrimentally impacting Florida's downstream use.[41] The allocations for Colorado River water have also experienced similar legal struggles.[42]

Disputant Interests

The practitioner should consider the parties' interests and anticipated outcomes as articulated during the preparatory phase. Accurately discerning and addressing the underlying needs of the disputants can be challenging, particularly when those needs are obscured by complex layers of symptomatic self-interest. To effectively navigate this process, disputants must independently reconsider their positions and, if necessary, reassess their bottom line or Best Alternative to a Negotiation Agreement ("BATNA"). [43]

The practitioner should seek to prevent individual interests from becoming the primary focus during the mediation. In the event that the competing interests over the resource continue, the focus remains on satisfying disputants' individual interests in order to end the mediation, rather than seeking to resolve the underlying causes of

---

[41] TOLUSE OLORUNNIPA & MICHAEL C. BENDER, *Florida to Sue Georgia in U.S. Supreme Court Over Water* (visited August 14, 2013), *at* http://www.bloomberg.com/news/2013-08-13/florida-to-suegeorgia-in-u-s-supreme-court-over-water.html; *See also* Florida's Motion for Leave to File a Complaint, Complaint, and Brief in Support of Motion (visited October 4, 2013), *at* http://www.eenews.net/assets/2013/10/02/document_gw_03.pdf.

[42] *See* SAN DIEGO COUNTY WATER AUTHORITY, *Water Authority Files New Lawsuit Against Metropolitan Water District Challenging 2013 and 2014 Water Rates* (visited March 12, 2013), *at* http://www.sdcwa.org/water-authority-files-new-lawsuit-against-metropolitan-water-districtchallenging-2013-and-2014-rates; *Colorado v. New Mexico*, 459 U.S. 176 (1982); LÉNA SALAMÉ, *ADR... What, when, where, who, how and most importantly WHY?!* (visited September 12, 2013), *at* http://es.slideshare.net/WaterforLife/mediation-and-dispute-resolution-techniques-and-approaches; YONA SHAMIR, *Alternative Dispute Resolution Approaches and Their Application*, Israel Center for Negotiation, and Mediation (visited February 4, 2013), *at* http://webworld.unesco.org/water/wwap/pccp/cd/pdf/negotiation_mediation_facilitation/alternative_di spute_resolution_approaches.pdf.

[43] *See* FISHER, *supra.*

the dispute.[44] This task is achievable provided the practitioner has laid the proper foundation by close examination of the objective criterion and how it relates to the issues at dispute.

After collecting sufficient information to form the CRB Model objective criterion, the practitioner must then create a simple presentation to explain this information and its implications to the disputants.

Conducting the Mediation and Reaching an Agreement

Facilitated mediation involves the engagement of practitioners who serve as impartial third parties to guide disputants through the mediation process. Facilitation aims to support the parties in developing solutions and reaching a mutually acceptable agreement, which is typically non-binding unless all parties consent to a binding resolution. The successful formation and implementation of such agreements are contingent upon the practitioner's expertise and proficiency.

Practitioners who have more active listening and questioning skills tend to be more effective throughout the mediation and successful in helping the parties find a solution. Practitioners with these types of skills provide disputants with a greater understanding of their opponent's interests by encouraging cooperation. Full disclosure by the practitioner at the beginning of the process is needed to establish an issue baseline and to start the trust-building process.[45] To instill these thought processes, the practitioner must facilitate open communication between the disputants to enhance understanding of true needs and separate the parties' perceptions of their interests. Cooperation can lead to better communication and trust, which thereby

---

[44] *See* COHEN, *supra*.

[45] *See* U.S. Fid. & Guar. Co. v. Dick Corp./Barton Malow, 215 F.R.D. 503 (W.D. Pa. 2003).

allows the disputants an opportunity to generate more individualized solutions for themselves. Unlike litigation, parties can directly communicate with one another and form mediation procedures. To achieve this challenging task, the disputants must make their own, more informed decisions, and re-evaluate their bottom line or BATNA, if required.[46]

To promote effective participation in mediation, disputants should be thoroughly informed about the potential benefits of the process. The practitioner is responsible for outlining the mediation process and the procedures for generating resolution options. It is essential to clearly articulate both the physical and legal frameworks to enable disputants to develop appropriate solutions that align with these parameters. According to prior agreements established during preparation, the practitioner will restate the guidelines and procedures for mediation to minimize confusion or escalation during negotiations. Additionally, the practitioner may share the analytical framework used to develop objective criteria, thereby reducing the likelihood of unnecessary or uninformed extended debate.

Mediation aims to encourage disputants to collaboratively generate and evaluate multiple resolution options, selecting the most suitable one for settlement. This process ensures thorough consideration of alternatives, increasing the likelihood of lasting success for all involved. Agreements should allow for verification and tracking of progress and resource availability. Practitioners guide parties through systematic evaluation steps, such as:

---

[46] Roger, F. and Ury, W., Getting to Yes, Negotiating Agreements without Giving In, Second Edition (1991) Penguin Group,

New York.

1. Outlining initial issues using methods such as mathematical matrices;
2. Identifying root causes and potential solutions for each issue; and
3. Assessing external influences on each option.

By presenting scenarios analytically, disputants can assign and rank values to options before making final selections based on objective criteria. The practitioner supports this process for each issue to help achieve balanced and sustainable outcomes.

Full disclosure and transparency by the practitioner at the beginning of the process are necessary to reestablish the disputed issues and initiate trust-building. At the commencement of negotiations, the practitioner must facilitate open communication between the disputants to enhance understanding of their true needs and to separate those needs from misperceptions of their opponent's interests. Cooperation can lead to improved communication and trust, allowing disputants the opportunity to generate more individualized solutions.

To further encourage participation during mediation, disputants must be fully informed of the benefits the process can provide. Based on procedural decisions made during preparation, the practitioner will reiterate mediation guidelines and procedures to prevent confusion or escalation during negotiations. The factors used to develop objective criteria must be clearly explained so disputants can feasibly generate resolution options. When appropriate, the analysis used in formulating these criteria may be presented to avoid unnecessary or uninformed protracted debate.

The goal during mediation is to facilitate collaboration and the creation of multiple resolution options. Once a variety of options has been generated, the practitioner may conduct a comprehensive walk-through of all dispute-related matters before the selection process begins by:

- Revisiting the initial issues identified during preparation;
- Identifying the symptoms and underlying causes of each issue; and
- Explaining how external forces, such as regulations, market conditions, perceived needs, and conservation concerns, may affect each option.

Ideally, the practitioner should educate and guide the disputants toward balanced solutions while simultaneously conserving the disputed resource.

Once the options are generated, disputants can evaluate and select the most appropriate option for implementation as a settlement agreement. Presenting scenarios in a matrix allows disputants to assign values to each option and effectively rank them. After narrowing the alternatives to a final few, disputants may request the practitioner to rank the options or recommend those that most closely adhere to the objective criteria. In practice, disputants should understand that this process may take more time than initially planned. Time is a critical factor at this stage of mediation, as constraints may pressure a disputant into agreeing to a settlement without full consideration, potentially resulting in unsuccessful implementation. Accordingly, the practitioner must ensure that adequate time is provided.

Once an agreement is reached, it must be constructed in a manner that allows for verification, if needed. However, disputants must be informed that mediation materials are often privileged and may not be admissible should a disputant later seek relief in court. Although solutions established through mediation are typically non-binding, disputants may elect to create an enforceable agreement. This may include signing a contract outlining each disputant's duties and referencing the mediated agreement. Disputants may also engage a third party to enforce or track the progress of the agreement, as well as monitor resource availability to encourage and incentivize compliance.

These mechanisms are examples of follow-up measures a practitioner may use to assess the success of an implemented agreement. Through continued analysis of such follow-ups, the practitioner can study successful outcomes and apply those methods or adjust tactics to ensure the CRB Model remains responsive to current demands and continues to provide long-lasting resolutions.

Chapter 4
Mapping the Environmental Regulatory Nexus

Environmental regulations establish a structured framework for developing solutions that equitably address the interests of all stakeholders while fostering responsible environmental management for the benefit of both current and future generations. These regulations delineate comprehensive standards that assist parties in navigating complex challenges and proactively mitigating legal and environmental risks.

The global commons refer to international, supranational, and worldwide resource domains where shared resources are located, including air, oceans, and minerals. These collective resources are subject to both potential and actual impacts from human activities, which may have direct and indirect consequences on a global scale. Even in the absence of formal regulatory connections between nations, actions related to natural resource consumption within one country may significantly affect the environments of others, thereby creating a regulatory nexus across national boundaries.

Natural resources, extracted from the Earth, serve as raw materials for manufacturing goods. While humans do not create these resources, they utilize and alter them to meet diverse needs. In the United States, regulatory oversight and permitting processes are distributed among federal, state, tribal, and local authorities. The issuance of environmental permits is contingent upon numerous factors, including the type of ore or source material involved, extraction methods, chemical storage protocols, and the nature of potential environmental discharges. Although most federal environmental permitting authorities are delegated to state agencies, certain exceptions remain. The primary objective of these regulations is to protect human health and the environment. Permitting agencies employ systematic evaluations of potential facility or mining operation discharges, including consideration of cumulative environmental impacts. A thorough understanding of both the intent and procedures of environmental and natural resource permitting enables practitioners to establish objective standards, resolve issues within the legal framework, and minimize prospective liabilities.

There is both a natural and legal connection between mediation and environmental issues, with the concept of circularity offering a framework through which technical natural resource information can be communicated and incentives created for parties involved in disputes. Multiple types of nexuses exist, some grounded in nature and others in law, and an understanding of resource life cycles and regulatory frameworks is essential. This chapter focuses on the legal nexus, specifically as it operates through established environmental regulations. The legal nexus functions as a bridge that facilitates cooperation and resolution among stakeholders and serves as a model for environmental and resource dispute management. Several critical considerations must be addressed to establish these legal nexuses:

1. Public Trust Doctrine
2. 1872 Mining Law
3. Water Rights and the Clean Water Act
4. Equitable Apportionment
5. Clean Air Act
6. Resource Conservation and Recovery Act
7. CERCLA
8. Superfund
9. Climate Change
10. National Environmental Protection Act

The Intersection of the CRB and Human Health

Most environmental regulations in the United States are designed to address public health concerns, forming the legal foundation for arguments focused on protecting both human health and the environment. The effectiveness of these regulations depends not only on their implementation but also on how stakeholders engage with them to achieve practical outcomes.

Over the past decade, I have closely followed the evolving landscape of environmental regulation and mediation practices. These frameworks often undergo significant shifts in response to changes in political leadership, rarely achieving prolonged equilibrium. Consequently, administrative transitions frequently result in substantial adjustments within the environmental regulatory context.

My professional background includes extensive experience interpreting regulatory standards and evaluating their implications for dispute resolution strategies. In mediations involving various natural resources, I have observed that many practitioners lack the specialized knowledge required to accurately interpret complex environmental regulations. Without this expertise, navigating the mediation process becomes significantly more challenging from a regulatory perspective.

There have been instances in which mediation outcomes inadvertently caused greater harm to the natural resources at issue. A central challenge for environmental professionals is effectively communicating to disputants the importance of resource conservation, emphasizing its dual role in environmental protection and in advancing the broader interests of all parties involved.

This overview examines major regulations and legal doctrines that commonly influence the resolution of environmental disputes. It is essential for practitioners to either possess, or assemble teams with, comprehensive expertise in the regulatory, economic, and technical aspects relevant to the resource under mediation. Such expertise is subject to continual scrutiny by all stakeholders, and the credibility of the practitioner, and their ability to facilitate equitable resolutions, depends directly on this knowledge base.

The following summary introduces several significant environmental doctrines and regulations. While it does not aim to provide exhaustive regulatory training, it seeks to identify areas where disputes may arise and to highlight certain regulatory incentives that may be available. Given the dynamic nature of case law and ongoing political developments, maintaining current and accurate knowledge of these issues remains critical.

Practitioners must discuss doctrines and regulations accurately, as disputants' perceptions of them directly influence mediation outcomes. Cultural and ideological connections to land shape legal trust, which is reflected in environmental legal frameworks. These frameworks link historical context with the mediation process and, through collaboration, help clarify issues and structure discussions.

Public Trust Doctrine

The public trust doctrine is believed to have originated in the sixth century A.D. within Roman civil law, which held that air, rivers, the sea, and seashores were owned by the state and dedicated to public use. Through various adaptations over time, the concept evolved into what is now known as the public trust doctrine, incorporating elements of French and Spanish civil law as well as English common law, all of which significantly influenced resource law in the United States. In the U.S., the public trust doctrine, shaped by its historical origins, represents a complex body of legal principles that evolved through a combination of federal and state law, including property, natural resource, and water law, among others. As early as 1892, the courts recognized the obligations and benefits of maintaining resources for all people[47]. In 1896, Supreme Court Justice White defined

[47] Illinois Central Railroad v. State of Illinois (1892) –form foundation for recognition of sovereign obligation for benefit of all people., United Plainsmen Assoc. v. North Dakota State Water Conservation Commission, 247 NW 2d 457 [1976], California

the issues of common ownership of natural resources and sovereign authority as the duty of the legislature for future generations[48].

Public trust is a trust that is created for a specific group, and in the case of public trust, for the general public, where a governmental agency is the trustee and the beneficiary is the public. In the case of the public trust doctrine, natural resources (assets) are placed under the control of a trustee (generally a governmental agency) for the public and future generations (beneficiaries). The public trust doctrine has several fiduciary duties that are encumbered by the construction of the document: a) the trustee has a duty to protect and preserve the natural resource (asset), b) a duty of loyalty to the beneficiaries, communication and disclosure of the resource to the public, c) a duty to be prudent in the delegation of authority[49].

The public trust doctrine obliges state and federal agencies to function as fiduciaries by balancing economic interests with beneficial uses. The Bureau of Land Management (BLM) is responsible for managing significant federal lands and is guided by its mission "to sustain the health, diversity, and productivity of the public lands for present and future generations." Disputes frequently arise from differing interpretations of the language within the BLM's mission statement.

An ongoing tension exists between conservation advocates and stakeholders pursuing land development for economic gain. Other resource or regulatory agencies with similar fiduciary duties under

---

Superior Court acknowledged the state's responsibility in consumptive water rights, in National Audubon Society v. Superior Court of Alpine County (189 Cal. Prtr. 346 [1983] check to see if it is still leading precedent

[48] Geer V. Connecticut, 161 U.S. 519, 529, 534 (1896)

[49] Courtney Lords, Protection of Public Trust Assets: Trustees' Duty of Loyalty in the Context of American Politics, Journal Envtl. Law and Litigation, Vol. 23, 519, 2008

the public trust doctrine include the United States Fish and Wildlife Service (USFWS) and the U.S. Forest Service (USFS). State resource agencies vary in their titles but typically include terms such as Natural Resources or State Parks, reflecting comparable fiduciary obligations. Many states, particularly in the western United States, have adopted their own versions of the public trust doctrine, each with distinct definitions and responsibilities. These interpretive differences are often at the core of recurring disputes.

Practitioners must carefully review agency mandates to determine the scope and application of their responsibilities. It is also essential to articulate how state and federal duties may intersect or, in some cases, create unforeseen obstacles. When addressing conflicts involving real property adjacent to public trust lands, practitioners should consider how proposed resolutions may affect those lands.

Because policies governing neighboring public trust properties can influence outcomes related to adjoining land, practitioners are advised to address these considerations early in the process. Scenarios that could escalate conflicts with neighboring properties should be avoided. Moreover, practitioners must clarify technical and legal terminology to minimize misunderstandings and prevent unintended complications.

Given the complexity of state and federal case law concerning the public trust doctrine, conflicts may arise from unintended barriers created by well-intentioned efforts to clarify issues or from inconsistent terminology that generates further disputes. Practitioners must remain attentive not only to these issues but also to how parties may strategically use such information in positioning themselves. Additionally, practitioners should be skilled at distinguishing among self-interest, political motivations, and fiduciary obligations.

Public trust interests are commonly understood in terms of economic or perceived value. Arguments regarding the doctrine's application vary widely. While many view the public trust doctrine as a tool for conservation, interpretations grounded primarily in economic interests may undermine public welfare, even when actions technically comply with state water allocation statutes or advance the political objectives of officials entrusted with fiduciary responsibilities.

Conflicts of interest may arise when the public trust intersects with an alleged breach of trustee duty. These situations can also result from divergent personal or political agendas and varying interpretations by governmental entities regarding beneficiary interests. Internal disagreements may occur within judicial agencies or among trustees whose personal political interests conflict with their professional responsibilities. State statutes generally require trustees to act in the best interests of beneficiaries, often encompassing all citizens of a state or nation, under what is known as the trustee's duty of loyalty.

State statutes should clearly define whether these duties include conservation and how terms such as conservation, reasonable use, preservation, and economic benefit are defined by law. Conflicts arise when parties interpret these terms according to their own perceived interests or needs. This divergence helps explain why interest-based processes alone can be exceedingly difficult to apply in achieving truly balanced resolutions, and why a CRB process may be more appropriate.

The practitioner's role is to guide parties toward replacing individual perspectives with a shared vision. During mediation, the practitioner should provide educational opportunities that enable participants to make informed decisions grounded in legal principles and resource constraints. This shared vision should emerge from the informational resources provided by the practitioner and be collaboratively developed by the disputing parties. Achieving this outcome requires comprehensive discussion of each party's interests and needs, as well as the underlying rationale informing their positions.

The 1872 Mining Law

The Mining Law of 1872 allows coal claims on federal lands; however, the Mineral Leasing Act of 1920 requires royalties for coal that is mined and sold. If a company develops a new coal mine to recover CREM, the operation could be classified as a coal mine, since coal must be extracted first. Permitting a new mine can be complex, often requiring an environmental impact statement that may take years to complete, as well as approvals to comply with the Surface Mining Control and Reclamation Act (SMCRA) and other federal laws. Mines must also obtain National Pollutant Discharge Elimination System (NPDES) permits if wastewater is discharged. Given the costly and time-intensive permitting process, new coal mines for CREM recovery are unlikely in the near term. However, existing coal mines may be able to amend their permits to include CREM recovery, and there are currently no regulations prohibiting this practice.

Water Rights and the Clean Water Act

Water is a finite resource, and access to clean, drinkable water is among the most critical challenges facing modern society. One need only examine the growing crisis in California to understand the limitations of freshwater availability. As a hydrologist, I can state that civilization will never run out of water on this planet; however, the availability of fresh, drinkable water is a separate and pressing concern. California, despite having over a thousand miles of Pacific Ocean coastline, continues to face severe shortages of potable water.

Fresh drinkable surface water is, in fact, the target of the Clean Water Act (CWA) in the United States. The CWA is based on the watershed approach and the hydrological cycle with associated principles within the watershed. The EPA website provides resources and training tools that the practitioner can use and that the disputants can read in advance to inform themselves of the watershed concept[50]. Understanding the concept of the watershed significantly enhances the practitioner's comprehension of how and why the Clean Water Act (CWA) applies to the protection of water resources.

The study of hydrology is relatively recent, emerging in the 1950s and 1960s as a synthesis of multiple scientific disciplines, including physics, chemistry, climatology, geology, soil science, and engineering. Hydrologists have gained a deeper understanding of how different hydrological processes are influenced by these disciplines, as well as the properties of water systems themselves. The ways in which these disciplines interact and co-evolve to address and manage water resource challenges provide a foundation for the CWA. It is therefore essential for the practitioner to understand this knowledge and convey it to disputants in a manner that supports the development of objective criteria.

Human interactions with natural resources are inherently complex and interconnected, with water playing a central role in transporting elements essential for life. Changes in water use can reduce availability for others downstream or in-stream, requiring responsible management to prevent overconsumption. These dynamics frequently lead to water disputes within and between states, typically driven by immediate needs rather than careful analysis of underlying causes or long-term solutions.

Protracted water disputes can take on a life of their own, often obscuring the central issues. Balancing interactions between disputing groups can be particularly challenging in elevated disputes, especially when there is no clear pathway to advancing understanding of resource limitations. Disputants may focus solely on immediate needs or

---

[50] http://water.epa.gov/type/watersheds/approach.cfm

strategically invoke water resource laws as bargaining tools for unrelated issues. Without advanced thought, disputes are likely to arise and remain for exceptionally long periods of time[51]. A systematic thought process is necessary to provide educational frameworks that demonstrate resource conservation. This approach often contrasts with the conventional thought patterns of most disputants. As disputes persist, the practitioner should emphasize that prolonged conflict harms both parties by reducing or eliminating access to the resource, increasing associated costs, and impacting future generations.

Thus, the practitioner's role as a mediator is crucial. Acting as a neutral third party, the practitioner educates disputants on the complexities and limitations of the resource and their respective roles in the dispute. The practitioner also facilitates communication to help each party understand the other's genuine needs, separating perceptions from underlying interests. At the same time, the mediator must convey the consequences of exceeding resource limitations under both man-made laws and natural laws. This is not an easy task, but it will allow the disputant to make their own more informed decisions, reevaluate their BATNA if required,[52] and achieve resolutions that not only benefit each disputant but also sustain the availability of the resource as well.

Equitable Apportionment

A review of current water doctrines in the United States reveals potential sources of conflict arising from unintended obstacles within legal frameworks and their mechanisms for achieving equitable relief. Mediating water rights disputes can be complex, particularly when parties adhere to differing legal doctrines. Such scenarios are

---

[51] Id. Sax & Abrams, 1986

[52] Id. Roger, F. and Ury, W., 1997

common in major interstate water conflicts, which are adjudicated in federal courts. To address challenges arising from conflicting state water laws, the U.S. Supreme Court may apply the doctrine of equitable apportionment to resolve interstate water allocation disputes. As a principle of federal common law, equitable apportionment guides the Court's decisions in these cases. For a state to prevail under this doctrine, it must provide compelling evidence of substantial harm, typically quantified in monetary terms. Demonstrating such harm can be challenging, however, because proof often requires that damage has already occurred, while case studies frequently rely on projections of future harm.

The high court may also select to use either prior-appropriate doctrine or riparian, or a combination of the two or neither when deciding a case. In the Kansas v. Colorado[53] case, where Kansas is a riparian state and Colorado a prior appropriation, the high court concluded that, where there are dulling laws or policies, equity would control the interest of the two states. In this case, Kansas claimed that Colorado's restriction of the Arkansas River under the riparian water law would harm Kansas' economic interests. The high court took a balanced approach based on economics, for which Colorado prevailed because Kansas could not meet the high hurdle of substantial harm, although the court did acknowledge the economic interest of Kansas.

In Wyoming v. Colorado,[54] where both states recognized the prior-appropriation doctrine, the high court refused to use the doctrine of prior appropriation. In the Connecticut v. Massachusetts case, where both states use the riparian doctrine, the high court refused to apply

---

[53] Kansas v. Colorado, 206 U.S. 46 (1907)

[54] Wyoming v. Colorado 259 U.S. 419 (1922)

the riparian rights doctrine and restated the "interstate common law" of equitable apportionment.

These cases provided the bases for future litigation, and in Nebraska v. Wyoming[55], a case that involved Nebraska, Wyoming, Colorado (impleaded), and the Bureau of Reclamation (intervener), where all three states were prior appropriation states, the high court again did not apply a strict prior-appropriations doctrine and instead applied the doctrine of equitable apportionment. In this case, the high court set out factors for a special master to consider regarding equitable water apportionment:

1. Physical and climatic conditions
2. Consumptive uses and the extent of established uses.
3. The rate of return flows
4. Availability of Storage water
5. The damage from upstream areas compared to the benefits to downstream areas if a limitation is imposed.
6. The practical effect of wasteful uses on the downstream
7. Overallocation of the resource

These factors are expected to form the foundation for future water disputes. Currently, equitable apportionment serves as a comprehensive framework not only for addressing water-related conflicts but also for disputes involving other natural resources. The Supreme Court has recognized unintended obstacles arising from existing state water doctrines and has, in some cases, rejected these approaches in favor of equitable apportionment. However, a notable limitation of equitable apportionment is its focus on economic considerations, often excluding ecological impacts.

---

[55] Nebraska v. Wyoming 325 U.S. 589 (1945)

Ecological aspects are challenging to quantify accurately using traditional economic models and do not fully account for constraints imposed by drought or climate change. This limitation underscores the argument that current processes allow for structural expendability of both resources and the communities that depend on them. For effective resolution of water disputes, practitioners should consider approaches that enable greater flexibility and adaptability in resource management.

Environmental practitioners are tasked with demonstrating and communicating the importance of objectively valuing ecological considerations. To support this, an overview of key environmental regulations is provided, along with a discussion of various funding programs that present opportunities for education and trust building. It is important to note that incentives associated with regulatory programs may not be available in cases involving violations. These incentives, offered as loans and grants, are designed to enhance the quality or availability of environmental resources. It is not the intent of the incentive programs to provide a benefit for those who have violated the regulations[56].

Clean Air Act

The Air Pollution Prevention and Control, or Clean Air Act (CAA),[57] is an oversight/preventative type regulation[58]. The CAA exemplifies federal collaboration with state regulatory agencies. Congress intended the act to protect public health from various forms

---

[56]http://www2.epa.gov/enforcement/water-enforcement-policyguidance-and-publications

[57] 42 U.S.C.A §§ 7401 to 7671q

[58] CAA § 101 (c)

of air pollution, which were particularly evident during the 1970s, the early years of the CAA's development. Major revisions were enacted in 1977 and again in 1990. The 13-year gap between these revisions illustrates the difficulty and controversy involved in updating the law and also reflects the challenges practitioners may face when navigating air regulations in a dispute. Positions in disputes often depend on each party's interpretation of the regulations and how they can leverage these interpretations to their advantage.

The objective of the CAA is to provide pathways for industry to transition from non-attainment to attainment areas. Practitioners can demonstrate the various pathways available for achieving attainment status. The CAA program also offers incentives in the form of credits, allowing industry to meet standards over time. Practitioners may apply this framework in disputes involving CAA incentive programs, which offer several options for meeting numeric standards through market-based incentives. As with all environmental regulations, these incentives may vary depending on the policies of the current administration. For the most accurate and up-to-date information, consulting official state and federal sources is recommended.

Resource Conservation and Recovery Act (RCRA)

RCRA is the primary federal law regulating hazardous and solid waste. Most states have delegated RCRA programs under USEPA oversight. Federal law allows states to adopt stricter regulations but not to reduce standards. Delegation levels vary, though most states have assumed full responsibility for implementing RCRA programs.

Unlike many broadly interpreted regulations, RCRA is highly enforceable. It establishes clear time frames, and exceeding them constitutes a violation. These provisions can clarify expectations for the mediation process, outline responsibilities, and help disputes progress through each stage. Practitioners should disclose all applicable regulations early and explain the consequences of noncompliance.

While water regulations focus on discharges to jurisdictional waters, RCRA regulates the disposal of waste into landfills and prohibits disposal at non-designated sites. RCRA addresses active and future sites, while CERCLA focuses on abandoned or historical sites. Both laws provide opportunities for CRB practitioners by emphasizing the cleanup of resources rather than individual perceived needs, unless the goal aligns with property remediation.

Brownfields

Brownfield cleanup projects offer financial incentives for remediating contaminated sites. The program supports environmental practitioners by providing grants, loans, and job training to revitalize polluted properties, especially in disadvantaged communities. Unlike other programs that rely on penalties, Brownfields emphasizes positive revitalization through site cleanup. Both approaches can be effective tools when applied constructively.

The Brownfields program enforces strict regulations for fund management and cleanup operations. It encourages collaboration and partnerships for properties that might otherwise remain idle. The program's focus on revitalization and cleanup aligns closely with the principles of the CRB mediation process.

Environmental and resource practitioners can leverage existing programs like Brownfields to provide funding and structure when working with municipal or industrial disputants. However, it is essential to introduce and apply these strategies at the right time and only where applicable to all parties. The program provides both structure and funding to encourage disputants when cleanup is at issue. There are hundreds of potential cleanup scenarios available, offering valuable tools for dispute resolution. These projects serve as trust-building, collaborative exercises that benefit disputants, local communities, employment, human health, and the environment.

Comprehensive Environmental Response, Compensation, and Liability Act (CERCLA)

CERCLA governs the cleanup of closed or abandoned hazardous waste sites by establishing liability rules, creating a trust fund when no responsible party is identified, and taxing the chemical and petroleum industries. The act sets procedures for identifying priority sites requiring removal or remediation and grants the EPA authority over private property and some federal lands. The EPA regulates hard rock mining contamination under CERCLA, while coal mine waste is managed by SMCRA and state agencies. Disputes related to CERCLA projects are typically addressed prior to site remediation and are managed according to CERCLA guidelines. The CERCLA site manager may implement Circularity concepts during remediation; however, CERCLA disputes may not always provide an ideal context for applying the CRB model.

Superfund

A brief historical look at the 1970s highlights two significant environmental disasters: Love Canal and Times Beach, Missouri. Love Canal was a hazardous waste disposal site where chemicals were dumped over several decades. In 1953, Hooker Chemical, then the owner, covered the pit and sold it to the city of Niagara Falls, New York, for $1. After record rainfall, the pit leaked and became one of the worst hazardous waste disasters in U.S. history, establishing the need for CERCLA and the Superfund cleanup program.

The Superfund program ensures long-term protection by overseeing hazardous waste removals and enforcing accountability for responsible parties. Both the Superfund and Brownfields programs have significant community outreach programs[59].

[59] http://www.epa.gov/superfund/community/index.htm

Engaging regulatory agencies early in the process helps align mediation efforts with state and federal requirements, reducing the need to revisit issues during mitigation and streamlining permitting. To build trust with environmental agencies, splitting samples can maintain quality control, promote transparency, and ensure correct procedural adherence. Early engagement enables project managers and practitioners to establish objective criteria using Circularity concepts for facility remediation.

Climate Change

The Clean Air Act (CAA) addresses climate change more directly than many other regulations, though approaches vary by administration. Climate change influences environmental doctrines and regulations, while global population growth increases resource consumption, heightening the likelihood of natural resource disputes. Because resources are limited and essential, improved problem-solving methods are crucial. Current mediation often fails by overlooking resource availability, leading to repeated disputes and inefficient outcomes.

Climate change produces direct effects such as extreme weather, indirect impacts like financial strain on insurers, and cumulative challenges that make proactive solutions increasingly necessary. Section 309 of the CAA authorizes NEPA, which establishes objective criteria for all NEPA project documents. The alternatives section of NEPA documents provides a structured guide for practitioners and stakeholders to assess options that comply with state and federal requirements.

Transparency and Decision Making

The NEPA decision-making process parallels the CRB process, offering a comprehensive framework that accounts for all relevant federal and state environmental regulations, though it does not issue permits itself. NEPA provides permit overviews and recommendations for applicants, guides alternatives analysis, and ensures transparent, inclusive decision-making that considers resource needs and stakeholder interests.

National Environmental Policy Act (NEPA)

For major NEPA projects in the U.S., the process is increasingly streamlined, with collaborative discussions between state and federal agencies to resolve differences and promote permit issuance during or before the Record of Decision (ROD). Revising NEPA regulations to advantage one side does not align with the original intent of NEPA's founders.

Not all projects require NEPA review. Four primary triggers can initiate the review:

1. Federal government land
2. Federal government permits
3. Federal government funding
4. An elevated level of controversy

The first three are the most common triggers; controversy is rarely invoked due to the expense of a full NEPA review. Most NEPA reviews concern federal projects rather than private ones.

NEPA establishes procedures to prevent environmentally harmful projects linked to federal funding or permits. Courts interpret and enforce NEPA regulations with guidance from the Council on Environmental Quality (CEQ). The process requires thorough evaluation of alternatives and affected areas for major actions with federal triggers, primarily involving federal money, land, or permits. Private projects generally remain exempt, except in rare cases of high controversy.

NEPA documentation can be costly and time-consuming, especially for contentious projects. Practitioners should consider these challenges when preparing an Environmental Impact Statement (EIS) and clarify who bears the associated costs. Courts apply a two-step approach to assess controversy: plaintiffs initially carry the burden of proof, followed by the agency's explanation if no controversies are deemed present. NEPA's analytical framework provides valuable guidance for resolving environmental disputes[60].

> *Sec.2. The purpose of this Act are: To declare a national policy which will encourage productivity and enjoyable harmony between man and his environment; to promote efforts which will prevent or eliminate damage to the environment and biosphere and stimulate the health and welfare of man; to enrich the understanding of the ecological systems and natural resources important to the Nation; and to establish a Council on Environmental Quality.*[61]

NEPA documents are, in fact, planning tools. Because they serve this purpose, they provide environmental practitioners with valuable opportunities. The planning process itself is inherently collaborative, particularly when managing large-scale projects that involve numerous complex issues. Successfully addressing these planning issues can determine not only the overall effectiveness of the planning process but also the success of the project. This approach parallels the analytical thought process used in selecting objective criteria for

---

[60] Foundation for North American Wild Sheep v. U.S Department of Agriculture

[61] The National Environmental Policy Act of 1969, as Amended.

environmental and resource disputes, a point I will revisit several times in this and upcoming chapters

Fig 4.1.

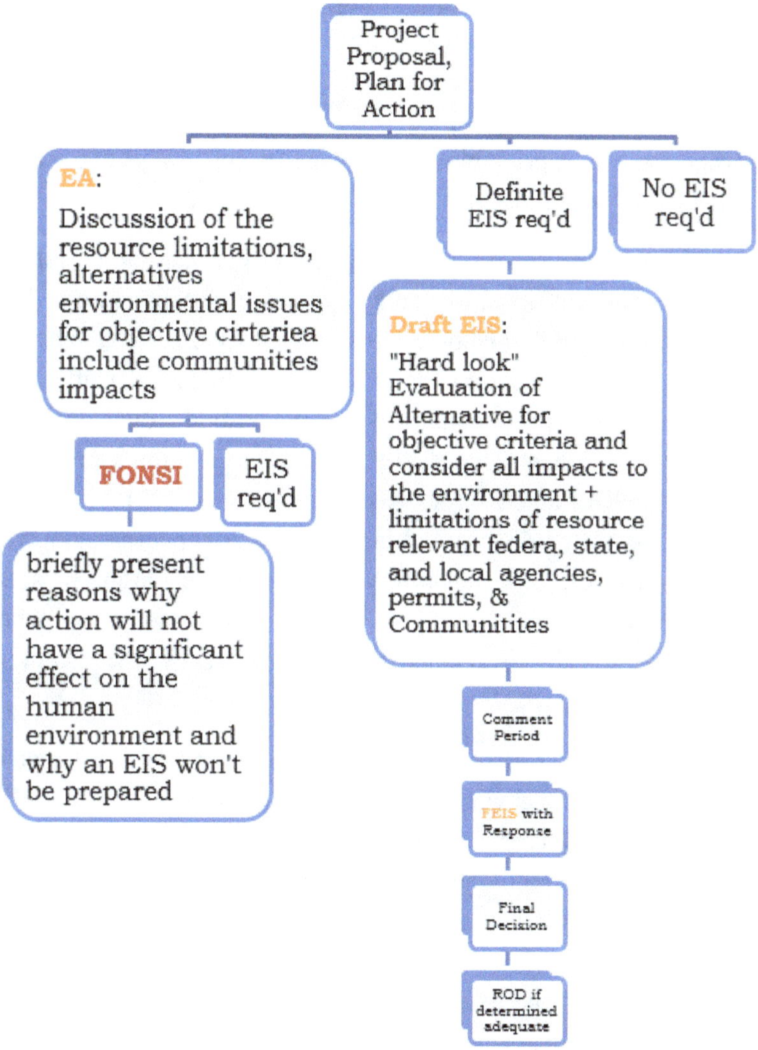

NEPA Procedures

The National Environmental Policy Act (NEPA), enacted in 1970, requires federal agencies to assess environmental impacts before making decisions. Agencies prepare Environmental Assessments (EAs) and Environmental Impact Statements (EISs) for projects that could significantly affect the environment. NEPA encourages informed decision-making and public participation. Understanding NEPA is crucial for grasping U.S. environmental policy and for leveraging NEPA principles in the development of objective criteria in natural resource disputes.

Natural resource disputes require decision-making at multiple stages during evaluation, often aligning with NEPA processes. These issues can be complicated by external influences or personal interests, which may cause practitioners to lose focus on central matters. To mitigate this, it is recommended that practitioners develop issue-specific decision-making outlines. NEPA documents provide a comprehensive framework for decisions involving natural resources, addressing both resource and environmental concerns alongside external factors. The standard NEPA evaluative outline presents issues logically and transparently, fostering trust among disputants. A systematic and logical evaluation process benefits the resolution of any dispute.

It is important to note that the current administration is undertaking revisions to several environmental regulations, including NEPA. This initiative, often described as "streamlining", is ongoing; however, the primary beneficiaries of these changes remain uncertain. Traditionally, the NEPA framework begins with a review of "Purpose and Need," which forms the basis for project analysis. Clearly defining the purpose in any dispute can illuminate the motivations of all parties, though this process may be complex. It is advisable to request comprehensive explanations from stakeholders regarding the origins of the dispute, as financial considerations are often pertinent.

The NEPA evaluation process systematically assesses project alternatives, ensuring the consistent inclusion of a "No Build" option. Although this alternative is rarely selected, feasibility studies may reveal that certain proposed actions are impractical or not cost-effective. Evaluating alternatives establishes objective and economically viable criteria. During this phase, stakeholders are encouraged to explore negotiable solutions while adhering to regulatory requirements, accounting for resource constraints, and assessing potential environmental impacts.

Next, the process examines the affected environment, with particular attention to how each alternative may influence the surrounding area. This stage involves analyzing the underlying causes of disputes and the roles of external actors. When utilizing the CRB process, practitioners address externalities and the dynamic interactions associated with human engagement in natural resources. Additionally, third-party liabilities are identified and assessed. Consistent with the NEPA framework, CRB evaluations require a thorough demonstration of the consequences of each alternative, including direct, indirect, and cumulative effects, to foster outcomes that are mutually advantageous for the environment, resources, disputing parties, and future stakeholders.

The analysis of consequences also incorporates restoration and mitigation considerations. Depending on the specific resource and nature of the dispute, these elements can serve as incentives or provide pathways toward agreement. Mitigation strategies may create new funding opportunities and stimulate discussions regarding credit systems.

The objective of evaluating all these issues simultaneously, while seeking alternative solutions that provide mutual gain, shifts the resolution away from personal self-interest and toward a sustainable outcome that benefits both the disputants and the resource in question. NEPA evaluative documents demand careful thought throughout the planning and decision-making processes. This in-depth, systematic and analytical type process creates effective decision tools for the practitioner for any type of mediation, even when natural resources are not involved. The alternative section of NEPA is truly the heart of any NEPA document[62]. This principle also applies when performing alternative analyses in natural resource mediations. The alternative analysis used in NEPA documents can function as a collaborative educational process, serving as part of the objective criteria employed in the CRB process.

Although not required for private projects with environmental concerns, practitioners can use this process to evaluate a project and establish objective criteria that all members of a dispute are likely to respect. In many cases, this approach also provides incentives through compliance.

Alternatives and Reasonable Solutions

Decision flowcharts and matrices are effective tools for illustrating the progression of decisions, highlighting which options were selected, excluded, and why. I prefer a spreadsheet matrix for issue evaluation because it allows practitioners and disputants to assign numeric values to proposed solutions where applicable. However, caution is warranted: if you have not previously worked within a numeric evaluation system, there is a risk that the numbers may come to overshadow the solutions themselves. Matrices work best with physical scientists and engineers who are comfortable with numbers. Numeric evaluation systems can be manipulated, potentially producing

---

[62] 40 CFR. §1502.14

results that do not accurately reflect reality. I always fall back on a quote from Mark Twain, “Figures don’t lie, but liars do figure.”[63]

The initial list of solutions serves as the foundation for creating alternative options, which can help establish the objective criteria used to shape potential resolutions. As each issue is thoroughly analyzed and solutions are proposed, these discussions build toward the necessary objective standards that should be incorporated into every proposed option. During this process, a preferred alternative begins to emerge and becomes clear to all participants. Typically, more than one preferred option may surface, and practitioners should anticipate that each party will develop an alternative reflecting their perspective on the issues. At this stage, the practitioner’s role is to guide everyone in reviewing all alternatives and distilling them into a single preferred solution. The challenge lies in ensuring that each party’s proposal accounts for any resource limitations related to the dispute. Once a preferred option is identified, all proposed alternatives are reviewed, impractical issues are removed, and additional factors, such as regulatory requirements, legal doctrines, resource limits, mitigation and restoration measures, human health considerations, and political and social contexts, are layered in. Ultimately, this process results in a preferred solution that addresses everyone’s needs and receives mutual agreement.

The Confluence of Legal Doctrine and Circularity

We are currently positioned at the intersection of legal doctrines and regulatory systems, a vantage point that allows us to systematically identify the central issues in disputes. Environmental law acts as a

[63] Mark Twain, 1913, who attributed the saying to the 19th-century British Prime Minister Benjamin Disraeli 1804–1881): "There are three kinds of lies: lies, damned lies, and statistics."

crucial link, connecting these concerns to real-world contexts and informing effective resolution strategies. By integrating environmental regulations with legal principles, objective standards can be established to create practical solutions for disputing parties. These frameworks provide practitioners with valuable tools for conducting mediation in a balanced and constructive manner while clarifying each party's strategic stance throughout the process.

The CRB Mediation system addresses resource limitations by treating them as variables within an integrated analytical framework, analogous to solving complex mathematical equations. Practitioners are advised to employ both evaluative and negotiated methodologies, using legal doctrines as constants to achieve balanced solutions. By explicitly listing variables and demonstrating how changes affect outcomes, practitioners can present objective scenarios for informed consideration. This modeling approach clarifies causal relationships and supports the thorough evaluation and presentation of resource constraints by integrating economic, scientific or engineering, and legal criteria into a unified matrix.

The development of objective standards grounded in credible scientific, engineering, and economic research ensures that all resolution options remain impartial and compliant with established benchmarks. These standards serve as guidelines for sustainable resource management within the confines of relevant legal frameworks. Criteria derived from scientific and engineering analyses, combined with pertinent market conditions, function as fundamental constants influencing outcomes, while legal frameworks define the permissible range of resolutions. Recognizing that change is the only constant emphasizes the need for adaptability throughout the assessment process.

Procedures for resolving water disputes vary significantly due to regional doctrines and economic considerations, which represent some of the most complex factors to evaluate. Although tools exist to estimate market value, few accurately capture the true cost of natural resources when externalities are included. Economic distortions, often stemming from government intervention, must be acknowledged by practitioners, as modeling may otherwise suggest that certain resources or communities are expendable. It is essential that disputants accept the models and tools used by practitioners, as these ultimately shape the outcomes of the resolution process.

Mediation practitioners routinely operate at the intersection of legal doctrine, policy, science, and economics, requiring expertise in managing these multidimensional factors to achieve sustainable and effective resolutions.

Fig 4.2

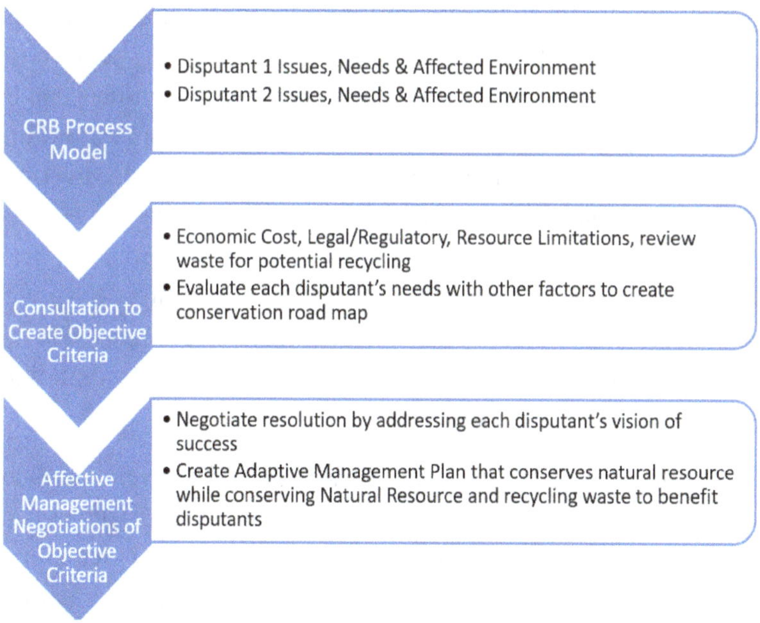

Chapter 5
Resource Economics Challenges and Opportunities

Parties involved in disputes often utilize a range of arguments to substantiate their positions, with influential stakeholders frequently relying on economic rationales to defend their actions or counter opposing perspectives. While economic decisions are commonly grounded in market valuations of resources, such assessments may not accurately represent true value. This disconnect can intensify conflicts and lead to increased expenditure of financial, temporal, and social resources.

This chapter introduces practitioners to various economic perspectives commonly found in environmental disputes and demonstrates collaborative methods, such as analyzing actual costs and building trust, for resolving them. Economic conflicts can be challenging but also present opportunities for effective solutions when approached with transparency. For example, resource use often raises employment concerns, as workers typically value job security. While environmental regulations are sometimes perceived as threats to jobs, it is possible to demonstrate that these measures, along with conservation efforts, can improve worker safety and longevity. In this way, employment can become an area where interests align with sustainable resource management.

Economics, as a mathematical field, provides several tools for addressing disputes. Presenting all viewpoints in economic discussions has historically shaped many important societal decisions. Environmental and resource mediators strive to achieve balanced outcomes in complex situations by collectively analyzing immediate priorities, such as employment, alongside broader objectives like protecting resources through economic strategies. Collaborative approaches often yield substantial benefits not only for local employment, taxpayers, and consumers, but also for society at large, both nationally and globally.

Natural resource trading has significant implications for both local economies and global communities. In 1832, Oxford political economist William Foster Lloyd examined pastureland grazing practices and determined that overgrazing, motivated by individual self-interest, resulted in diminished economic returns. Farmers increased livestock numbers to maximize profit, but this ultimately reduced cattle value and compromised the pasture's long-term productivity.

In this context, the resource itself becomes the limiting factor. Absent sustainable management, negative financial outcomes and resource scarcity affect farmers, consumers, and future generations alike. As supply declines, consumers encounter higher meat prices.

This case exemplifies how decisions at the community level can influence wider markets. Many economists and environmental professionals regard this scenario as the foundation of what is now termed the "Tragedy of the Commons" (TOC). It is critical for practitioners to understand the economic dimensions of environmental and resource-related disputes, recognizing that effective solutions must account for both immediate community needs and broader regional or global dynamics. The TOC framework is relevant to issues such as forestry, water pollution, fisheries, climate change, and banking.

Economic arguments are often used to oppose environmental regulatory measures, suggesting that such regulations lead to job losses and increased business expenditures. However, an alternative viewpoint emphasizes the ways in which regulations can support public health, protect environmental resources, and contribute to employment generation.

The practices of outsourcing government services and reducing agency budgets have become increasingly prevalent as strategies to decrease expenditures and minimize government size. Nevertheless, this approach can inadvertently overlook the costs associated with maintaining critical infrastructure, such as water and wastewater treatment facilities. Historically, advancements in these sectors have contributed significantly to public health and environmental quality. Despite their essential role, these facilities frequently become targets for budget reductions at the local government level.

Water and wastewater treatment operations not only support public welfare but also generate valuable employment opportunities and provide social and economic benefits to local communities. Small rural communities often face challenges in attracting and retaining the specialized expertise required for effective facility management. Contracting with management firms represents one potential solution, wherein external entities oversee daily operations while municipal authorities retain liability. Outsourcing may offer cost savings under certain circumstances; however, excessive costs resulting from such arrangements may ultimately be passed on to taxpayers. It is important to recognize that businesses engaged in service provision are primarily profit-driven, whereas government agencies are fundamentally responsible for addressing public needs.

Supply & Demand and Cost-Benefit Analysis

Economic considerations are integral to the resolution of environmental and resource disputes. Analysis typically begins with identifying all associated costs and benefits, followed by assigning valuations, often expressed in monetary terms, to each category. Accurately valuing resources presents significant challenges; when true worth is not reflected and resources are regarded merely as commodities, cost-benefit analyses may become distorted.

A widely adopted tool is the supply and demand (S&D) curve, which focuses solely on variables of supply and demand without accounting for the intrinsic value of resources or the long-term costs of depletion for future generations. Historically, certain administrations have relied on variations of supply-side economics, at times neglecting conservation priorities. Well-documented examples include the 1980s Savings and Loan bailout and more recent banking crises, in which asymmetrical liability distribution placed burdens on taxpayers and exposed shortcomings in specific valuation methodologies.

To safeguard the interests of future generations, it is essential to develop balanced frameworks that equitably distribute liabilities and prioritize resource conservation. Concepts such as the Tragedy of the Commons (TOC) and supply and demand illustrate how insufficient attention to conservation can generate disproportionate liabilities and losses across local, regional, national, and global contexts.

Natural resource conflicts should no longer be viewed as purely local concerns. Economic arguments significantly influence power dynamics within disputes and remain prevalent in regulatory and policy arenas. Perspectives grounded exclusively in supply and demand reveal notable limitations, as highlighted by the "Tragedy of the Commons," which addresses individual benefit and self-interest.

Although interest-based systems aim to separate personal motives from individuals, prevailing economic arguments and legal structures often allow parties to reintroduce self-interest under economic pretenses. As a result, parties relying heavily on economic reasoning frequently prioritize immediate gains while externalizing costs, which are commonly borne by third parties such as taxpayers. Taxpayers are often responsible for funding environmental remediation following pollution or for addressing resource shortages that require imports and may increase consumer costs.

Since the early twentieth century, increased fossil fuel consumption and industrialization have significantly influenced both environmental conditions and standards of living. While technological advancements have enhanced convenience, they have also introduced notable implications for human health and the environment. Historical data reveal that changes in air and water quality have lasting effects on public health. Measures of human health are frequently employed to assess the environmental outcomes of industrialization, and government intervention in pollution-related issues is often informed by factors such as healthcare costs. Progress in regulatory efforts can vary depending on shifts in political leadership.

Cost-benefit analysis remains a central method for evaluating regulations and projects, with job creation regularly cited as a primary benefit. Both costs and benefits are typically appraised in monetary terms; however, assigning market value to natural resources presents challenges, particularly when accounting for future expenses influenced by subsidies and technological change. As a result, the methodology used in cost-benefit analysis can become a significant source of contention. Professionals must therefore be aware of how stakeholders leverage market-based strategies in resource negotiations.

A fundamental principle established in physics, the "Conservation of Mass," asserts that mass within an isolated system remains constant. In contrast to closed systems, natural resources operate within interconnected cycles. For example, Earth's finite water supply persists through the hydrologic cycle, underscoring the interrelationships among water processes. Modern water treatment technologies are often modeled after these natural purification mechanisms. The costs associated with the hydrologic cycle extend beyond traditional economic metrics, and failure to account for these costs may compromise effective resource management. In this way, the hydrologic cycle illustrates the complex interplay between natural systems and dispute resolution frameworks.

When employing the interest-based approach, certain environmental and resource-related costs may be overlooked, which has prompted proposals for alternative frameworks such as the CRB system. The interest-based model may not consistently achieve equitable outcomes in natural resource and environmental conflicts, as it can prioritize individual advantages over collective needs. Without careful integration by experienced practitioners, these approaches risk neglecting conservation concerns and the fair allocation of responsibilities among current and future stakeholders. Economic models that exclude conservation considerations can further impede effective mediation by limiting opportunities for sustainable solutions, and zero-sum tactics are often prevalent within these methodologies.

Zero-Sum

The zero-sum approach, which aligns with certain negotiation styles, can create challenges by framing disputes as situations in which one party's gain is another's loss. In resource negotiations, the most influential party may secure a disproportionate share. Addressing zero-sum issues requires identifying areas of shared interest, often through consensus-building and educational techniques. Informational exchanges, such as modeling, stakeholder interviews, and surveys, can help participants better understand resource limitations and explore solutions that support both present and future needs. Environmental practitioners play a key role in facilitating equitable distribution and representing resource interests impartially.

The influence of natural resources in disputes largely depends on the methodology used in cost analysis. In many cases, resources themselves do not have an active role, and success is measured by allocation rather than sustainability. Economic models can project monetary value based on availability, with resource depletion typically leading to higher valuations. Educational initiatives are therefore critical before depletion occurs. Externalities, which may not be reflected in market prices, often influence resource use and depletion. Regulations are intended to address these externalities and mitigate adverse impacts, raising questions about how to balance immediate human benefits with obligations to future generations and how to minimize disproportionate effects on vulnerable groups. Regulation contributes to balanced outcomes by establishing controls over resource access and use.

Market Forces

Market-based systems seek to regulate demand and resource use through mechanisms such as rate structures, fees, taxes, and tradable permits. Each approach has advantages and disadvantages and may be perceived differently depending on economic position. For instance, increasing rates may have minimal impact on wealthier individuals while disproportionately affecting others. Taxes and fees can function as incentives or disincentives; however, accurately determining appropriate levels for environmental protection is complex and often constrained by limited data. Public acceptance of new taxes is typically low, even when they are intended to provide community benefits. Practitioners must therefore carefully evaluate when to recommend market-based systems.

Several market tools establish systems designed to reduce pollution or resource consumption by linking costs to usage. These mechanisms encourage industry to invest in mitigation when doing so becomes more economical than continued pollution. Nevertheless, criticisms persist regarding their effectiveness and fairness, particularly in relation to externalities and lobbying influences. Market approaches remain contested in terms of their ability to control overconsumption and pollution, with some arguing that they primarily enable third-party profits. Practitioners should consider these concerns to ensure that dispute resolution processes are both effective and equitable.

Practitioners must be able to identify when, and if, these techniques provide balanced solutions rather than serving merely as revenue sources for third parties or governing agencies. In some cases, all of these practices may be appropriate; in others, none may be suitable to achieve balance in the process.

Market Incentives

Market incentives within environmental regulatory programs can offer opportunities for stakeholders, provided they comply with regulations, collaborate with local agencies, and meet program requirements, which often involve infrastructure projects.

Common market incentives include subsidies, fee systems, and marketable rights. Subsidies, funded by federal or state sources through tax deductions, grants, or low-interest loans, frequently cover a substantial portion of project costs, with matching funds often required. Local subsidies have reduced drinking water costs for many years, but can also contribute to accelerated resource depletion.

While most market incentives focus on resource use, some programs require renewable resources to be replenished and seek to stabilize market prices, though they rarely reduce overall demand. In certain cases, incentives may even increase demand if consumers perceive resources as abundant and inexpensive. Nonetheless, sustainability-oriented incentives are increasingly promoting alternatives to existing practices.

Information Disclosure

The concept behind information disclosure is that public disclosure reduces the benefit of polluting or consuming unnecessary natural resources in a non-beneficial way. Public disclosure through the Emergency Planning and Community Right-to-know Act[64] (EPCRA) and the National Environmental Policy Act[65] (NEPA) are two examples of information disclosure approaches. The assumption is that public disclosure will result in the public responding to issues, and thereby the user may elect to select a less controversial process or protect the natural resources in question. It should be noted that there are no guarantees with this process because the public may not elect to become engaged in the process; however, exposing potential third-party liabilities may help the parties to a dispute reconsider their position.

The CWA provides citizen suits[66], in that the public may sue the polluter or regulatory agency for violations of the CWA. The logistics and costs associated with these suits were believed to provide deterrents to violators, as they might discourage individuals from polluting. However, the cost of bringing a citizen suit often has a greater deterrent effect on the public than on violators themselves, due to the expenses involved.

Most state and federal regulations include some form of public disclosure requirements; however, how effective these disclosures are in deterring pollution is unclear, as I am not aware of any statistical studies that definitively prove or disprove their effectiveness. Public disclosure is seldom, if ever, a method employed by mediation practitioners. Several challenges are associated with incorporating public disclosure into the mediation process. This does not mean that public disclosure cannot be used, only that if a practitioner elects to employ it as a tool, they should be aware of the potential legal liabilities, as discussed in previous chapters.

---

[64] 40 C.F.R. Part 302, 350, 355, 370, 372, and 374

[65] 40 CFR 1506.11

[66] 33 U.S.C. § 1365

Resource Credits/Banking

The primary objective of resource credit and banking systems is to create a market-driven approach that establishes an overall cap on industry-wide pollution emissions, assigning costs to discharges from industries that fail to meet new standards. This mechanism, commonly referred to as "Cap & Trade," was initially introduced under the Clean Air Act (CAA), with the Acid Rain Program serving as its first application to reduce SO_2 emissions. Cap & Trade programs allow industries to determine the most effective methods for reducing emissions while permitting the purchase and exchange of credits to achieve regulatory compliance. These programs are typically structured with a transitional period, providing industries time to adapt before full compliance is enforced.

Although Cap & Trade was originally conceived as an alternative to traditional Command and Control regulation, in which agencies set explicit emission limits, regulatory bodies continue to determine the cap and enforce compliance. Notably, the expectation that Command and Control mechanisms would resume following initial compliance periods has not always been realized.

Practitioners encounter multiple complexities when incorporating Cap & Trade mechanisms into resource management frameworks. A key challenge involves aligning pricing with allocation quantities; affluent entities may delay investments in pollution mitigation by purchasing credits or negotiating preferential regulatory terms, thereby prolonging periods of elevated emissions. Additionally, market volatility introduces risks, including the potential for unethical practices. Uncertainty also remains regarding the achievement of overall industry emission targets, an issue similarly associated with emission tax models. Although taxes can generate revenue for regulatory agencies, industries may choose to pay these fees until they are able to negotiate more favorable conditions. Vigilance is therefore essential to identify and address strategic behavior while maintaining safeguards designed to promote equity within the system. When administered equitably, Cap & Trade represents an innovative approach to environmental regulation.

This market-driven strategy has been adopted across a range of environmental and resource sectors. The relative effectiveness of Cap & Trade initiatives has supported their application in areas extending from air quality management to local water governance. In regions of the western United States where water scarcity is acute, regulatory authorities frequently employ tiered rate structures that increase costs as consumption rises, thereby encouraging conservation. Some areas also mandate multiple cycles of water reuse by consumers. However, differing interpretations of "reasonable use" have contributed to continued overuse, particularly among individuals capable of absorbing higher costs. Practices such as maintaining ornamental fountains and waterfalls during drought conditions may comply with legal standards, yet they raise questions regarding prevailing assumptions about prudent resource use.

Process Design

Effective mediation of environmental and resource-related disputes depends on rigorous process design. The structure and adaptability of the mediation framework are fundamental to achieving successful outcomes. Prior to engaging stakeholders, practitioners should conduct thorough research into the disputed resources, the relevant political landscape, and the historical context of the conflict. Detailed plans should then be developed to address identified issues and outline corresponding strategies.

An adaptive management outline serves as a strategic guide, helping parties navigate potential ethical and legal challenges. Reliance on inflexible or generic models can exacerbate conflicts or inadvertently violate regulatory requirements. Importantly, process design must align with the specific nature of the dispute; complex, multi-party resource conflicts rarely conform neatly to either interest-based or rights-based paradigms. Historical experience demonstrates that inadequate analysis and the absence of a shared vision can undermine sustainable resolutions.

A common challenge arising from poorly constructed outlines occurs when practitioners themselves become focal points of blame, particularly if disputants fail to make progress or engage in prolonged verbal confrontations. In such cases, responsibility may shift toward the mediator. Consider a scenario in which a mediator also serves as a moderator, such as a police officer responding to a domestic dispute. If the officer is adequately trained and follows established protocols, escalation may be minimized; nevertheless, the unpredictable nature of domestic disputes underscores the importance of standardized procedures. Officers typically separate involved parties to de-escalate tensions and gather individual perspectives through targeted questioning, thereby enabling informed decision-making. This approach alters the dynamics and atmosphere that initially produced the conflict. In straightforward cases without violence, resolution may be swift and satisfactory. However, if disputants reject recommendations, the practitioner may be perceived as the source of the conflict.

It is not uncommon for disputing parties to view the practitioner as an unwelcome external presence and redirect their frustration toward that individual or their team. As illustrated in the example of the police officer, parties may shift focus from their disagreement to collective criticism of the practitioner. Accordingly, comprehensive training and well-defined protocols provide practitioners with essential negotiation tools, enabling them to respond effectively based on both stakeholder behavior and procedural guidance.

An inadequately planned mediation may result in disputing parties redirecting aggression or assigning blame to the practitioner or their team members. This scenario illustrates a common pitfall in mediation: the necessity of establishing and adhering to a comprehensive plan. In such circumstances, the practitioner might be viewed as responsible for a party's frustration if the dispute is not resolved in a way that aligns with their preferences. To mitigate this risk, it is essential for the practitioner to provide the disputants with a clear outline of procedures and expectations early in the process, ensuring that all parties understand their roles and the standards of productive conduct required. Liability for not following state guidelines is increasing and differs from state to state[67].

Strategy

A practitioner's primary approach should be to clarify and illustrate both legal and natural connections, demonstrating how these relationships can benefit everyone involved. Once established, these connections act as a *knowledge bridge*, helping disputants gain clearer insight into the issues so they can develop their own solutions using information generated through mediation. The more thorough and relevant the information shared, the more effectively parties can engage with one another. When a dispute involves uneven levels of knowledge between parties, however, unique challenges arise. Typically, I begin by assuming that all participants understand the core issues; nevertheless, to ensure clarity, I start with the fundamentals and gradually introduce more complex topics, adapting the process to each party's comfort level.

---

[67] See Uniform Mediation Act and applicable state statutes, and http://www.americanbar.org/groups/dispute_resolution/policy_standards.html

Creating a flexible standard procedure for environmental or resource disputes may be the most effective starting point for practitioners. While not every participant in environmental mediation arrives with a defined strategy, most do; those without one may appear disorganized or difficult. Practitioners frequently encounter challenging individuals, but those who can discern why certain parties present difficulties gain a significant advantage. Truly understanding the parties requires recognizing and identifying the strategies they employ.

It is understood that most people's strategy is based on their bottom line or their BATNA[68]. What we must remember is that for every action there is an equal and opposite reaction, and for every advantage there is a corresponding disadvantage. Accordingly, a BATNA that provides an advantage to one party often forces the opposing party to respond with an opposing reaction, generally resulting in disadvantages for both sides. When parties are unwilling to adjust their BATNA, the process can become gridlocked. Disputants who are unorganized or labeled as difficult present additional challenges for the practitioner. Simply asking a disputant to identify their bottom line can initiate dialogue and help move the process forward. It should be noted that disputants may initially resist stating their BATNA, but they may become more willing once it is clear that disclosure serves their interests. Difficulties arise when parties refuse to deviate from their bottom line, decline to share it, or are uncertain what their bottom line actually is.

Confidentiality

Confidentiality is often subject to misinterpretation in the context of mediation. While some practitioners regard it as an infallible safeguard,

---

[68] Id. Fisher & Ury (1991)

others view it as a potential liability; nevertheless, its fundamental importance lies in its role as a vital component of the mediation process. Participation by disputants would be minimal without assurances that their concerns and disclosed information are protected. The central function of confidentiality is to promote candid communication within a neutral setting. For practitioners, upholding confidentiality means earning the trust of the parties by managing sensitive information with discretion throughout the mediation.

In environmental disputes, where disclosure of business practices or industry data may be necessary, safeguarding such information is imperative unless superseded by state or federal law. Matters involving industrial operations, trade secrets, or proprietary data that could influence competitive positioning are classified as confidential business information (CBI). Such information often arises unexpectedly during mediation, requiring a strong foundation of mutual trust among participants. Confidentiality or non-disclosure agreements may be implemented to protect trade secrets and other CBI, and these agreements are typically enforceable under contract law.

A range of CBI issues can surface in environmental and resource-related conflicts. For example, mining disputes may involve information regarding ore volumes or the locations of historical finds. Practitioners must remain aware of applicable legal frameworks to ensure such information remains protected, particularly when disclosure could confer a competitive advantage or increase the risk of post-mediation litigation.

Confidentiality stands as a cornerstone of effective mediation, facilitating comprehensive discussion of contentious topics. It protects both the integrity of the process and the interests of the parties involved, while also posing risks to practitioners who fail to adhere to statutory standards. Most jurisdictions establish clear expectations for professional conduct in mediation practice.

Developing a thorough outline of the mediation process helps practitioners avoid ethical and legal errors. Incorporating established legal and ethical guidelines into this framework, and informing participants of procedural protocols, enhances transparency and reduces the likelihood of post-resolution disputes. Careful process design is essential to successful mediation outcomes; insufficient planning can undermine confidence and increase mediator liability, ultimately affecting the profession's reputation. These pitfalls most often occur when parties inadvertently violate ethical standards or breach confidentiality obligations.

Practitioners working in jurisdictions that require disclosure of information related to child abuse or illegal drug activity must comply with such mandates, even though these issues rarely arise in environmental disputes. Nonetheless, matters involving CBI and resource management frequently surface. While practitioners often rely on confidentiality as a protective measure, it is not always a valid legal defense. Environmental mediations typically occur in closed sessions; however, when government agencies or public interests, such as parks or public lands, are involved, open meeting laws must be incorporated into the process design to ensure compliance with applicable legal requirements.

Neutrality

Neutrality is a fundamental element of mediation and requires mediators to remain impartial toward all parties. Mediators must avoid conflicts of interest related to the outcome of the dispute. Although neutrality appears straightforward in theory, it becomes more complex in practice due to shifting interests and multifaceted issues.

In formal Alternative Dispute Resolution (ADR) settings, such as arbitration, arbitrators, who may be attorneys, judges, or subject-matter experts appointed by a court, render decisions grounded in law rather than advocating for either party. In certain circumstances, an arbitrator may advocate for specific interests, such as those of children in family law cases, when permitted by legal standards and without compromising impartiality.

Informal ADR processes require mediators to maintain objectivity by refraining from taking sides or offering advice to disputants. An effective mediator preserves independence, avoids personal relationships with participants, and remains unbiased regarding the subject matter. In some instances, environmental practitioners may support resource conservation measures without violating neutrality; for example, advocating for financial assurance in mining operations ensures proper restoration and monitoring, ultimately benefiting all stakeholders.

Practitioners can also address community concerns through conflict prevention and by strengthening long-term agreements following mediation. When outside interests are represented, practitioners must strictly adhere to ethical standards to maintain impartiality.

Concerns regarding practitioner behavior may arise, making protective measures such as liability insurance and Errors and Omissions coverage necessary. However, insurance cannot substitute for well-designed mediation procedures. Quasi-judicial immunity may apply in certain circumstances, but its scope varies by jurisdiction and is typically limited to court-ordered mediations. Ultimately, practitioners must understand how liabilities related to neutrality shift depending on the type and setting of mediation in order to effectively navigate their responsibilities.

Beyond Neutrality

In most situations, practitioners are expected to maintain neutrality and avoid influencing outcomes in favor of either party. Nonetheless, practitioners employ a range of facilitation strategies to support dispute resolution. When one party is uncooperative, practitioners may implement measures to encourage participation; however, if not carefully managed, such interventions can inadvertently be perceived as advocacy and raise legal or ethical concerns.

Some professionals argue for an expanded role that moves beyond strict neutrality to offer a broader array of services to parties in conflict. While supporting a particular perspective carries inherent risks, environmental practitioners may appropriately promote resource conservation when such efforts benefit all stakeholders and ensure the continued availability of those resources.

The adoption of a conflict engagement specialist approach involves educating parties about the immediate crisis, particularly regarding limitations on natural resources, and guiding them through a structured analytical process. This approach enables collaborative identification of solutions that serve the interests of the parties, the environment, and the affected resources.

Environmental practitioners are therefore encouraged to prioritize constructive conflict engagement over resolutions that address only the immediate interests of disputing parties. Through this shift, practitioners can make meaningful contributions to advancing the field of environmental mediation and redefining effective conservation practices.

Ethics

Ethics extends beyond professional regulations, serving as a philosophical framework grounded in natural rights and legal standards across most professions, including environmental practice. Licensing and ethics boards establish codes of conduct specific to each discipline, though underlying moral principles are often shared. Mediators operate within social contexts shaped by cultural, religious, and legal values and must foster understanding among parties with differing perspectives. Clear communication and early consensus-building are essential to effective mediation.

Ethical standards also vary by profession. Attorneys adhere to rules established by bar associations, while scientists and engineers are regulated by state licensing bodies. In many states, mediators are required to be certified, typically through a minimum of 30 hours of formal training.

Power and Position

Power is described by practitioners as "...the currency of conflict...[69]" In this context, power denotes the ability to fulfill one's needs and advance objectives. Disputants often seek to leverage positioning to gain strategic advantages during negotiations, particularly in environmental or resource-related conflicts. Power in these negotiations can arise from several sources:

1. **Formal authority**, granted through laws or regulations.
2. **Informational power**, derived from knowledge of the disputants and an understanding of natural resource limitations. Such knowledge enables practitioners to facilitate equitable outcomes that address stakeholder needs while supporting long-term sustainability.
3. **Associative power**, whereby communities lacking political or financial influence align with larger entities, such as NGOs, governments, or industry partners, to strengthen their negotiating position.
4. **Resource-based power**, which is especially relevant in low-income or minority communities with limited

[69] Beyond Neutrality, B.S. Mayer,2004

economic resources. In these cases, economic valuation tools may be used to elevate or diminish the perceived worth of resources or communities, sometimes resulting in structural expendability.

Market mechanisms, including regulations, disclosure requirements, and cost-benefit analyses, are commonly used to establish economic positions. However, these tools can undervalue natural resources or affected communities when applied narrowly. Misuse may reflect systemic bias and overlook external costs borne by vulnerable populations. A balanced methodology that emphasizes true resource value is therefore essential.

As Shellenberger and Nordhaus observe, a narrow definition of self-interest within the environmental community can impede effective policy outcomes. Their critique highlights the risks of rigid, literalist policy approaches that may undermine constructive conflict resolution.

Positioning remains a prevalent negotiation tactic, with parties seeking to maximize perceived advantages through legal authority, zero-sum frameworks, market incentives, information control, and evaluative tools such as cost-benefit analysis. Recognizing the inherent limitations and risks of these strategies is critical for practitioners throughout the mediation process.

Mayer further categorizes negotiation power into four forms: (1) formal legal authority, such as the Civil Rights Act or Executive Order 12898; (2) informational power, rooted in comprehensive understanding of both disputants and resource constraints; (3) associative power, as politically marginalized groups form alliances to amplify their influence; and (4) resource power, reflected in the political and economic significance attributed to resources themselves. Market tools are frequently misapplied to justify unfavorable outcomes for minority communities, highlighting the complex intersection of science, law, and policy.

Structural expendability disproportionately impacts minority and Environmental Justice communities, which are often burdened with environmental costs. Emergy-based approaches may provide more balanced assessments of resource value, yet structural expendability has historically been used to rationalize the placement of hazardous activities in vulnerable areas. Practitioners must identify these dynamics early in the research process to support more equitable negotiations.

Repeated reliance on litigation or mediation often signals unresolved or unsustainable outcomes and can erode trust among parties. Practitioners should therefore emphasize balance and issue interconnections, ensuring that settlements address disputant needs without compromising ecological stability. Integrating market forces with ecosystem cycles and giving due consideration to alternative solutions supports long-term resource preservation and strengthens the overall mediation process.

Bibliography

Beer, J.E., Stief, E., The Mediator's Handbook, third edition, New Society Publishers, (1997)

Black' Law Dictionary, 7th Ed. West Group (1999).

Burgess G.M., Current Challenges in Environmental Dispute Resolution

www.colorado.edu/conflict/full_textsearch/ALLDocs/89-6.htm

Cohen, Stewart, Waddell, Melissa, Climate Change in the twenty first Century, Kingston, ON, Canada, McGill-Queens University Press, (2009)

Coleman, P.T, Fisher-Yoshida, B., Stover, M.A., Hacking, A.G. et al., (2008) Reconstructing Ripeness II: Models and Methods for Fostering Constructive Stakeholder Engagement Across Protracted Divides, 26 Conflict Resol. Q. 1.

Bartoli, A, Reconstructing Ripeness II: Methods for Fostering Constructive Stakeholder Engagement Across Protracted Divides, Conflict Resolution Quarterly, vol.26, no. 1, Wiely Periodicals, INC. (2008)

De Young, R. (1999) Tragedy of the Commons. In D. E. Alexander & R. W. Fairbridge [Eds] Encyclopedia of Environmental Science. (Pp. 601-602) Hingham, MA: Kluwer Academic Publishers

Dunne, L. Leopold, (1978) Water in Environmental Planning, W.H. Freeman & Company

Evans, A.M. and Kreueger, J.I. (2009), The Psychology (and Economics) of Trust. Social and Personality Psychology, compass, 3:1003-1017. doi: 10.1111/j. 1751-9004.00232

Fisher and W. Ury, (1983) getting to Yes: Negotiating Agreement Without Giving In, (New Yourk: Penquin Books, 1983)

Gartenstein-Ross, Daveed, An Analysis of Rights-Based Justification for Federal Intervention in Environmental Regulation, Duke Environmental law & Policy Forum, Vol. 14, No.1, 2003

Gerrard, Michael, and Foster, Sheila, (2008) The Law of Environmental Justice: Theories and procedures to Address Disproportionate Risks, ABA Publishing, Chicago

Gerzon, Mark, Leading Through Conflict Harvard Business School Press, Boston Massachusetts, (2006)

Houch, Oliver. Tales from a Troubled Marriage: Science and Law in Environmental Policy, Tulane Law School, www.sciencemag.org/content/302/5652/1926

Kalen, S., Ecology Comes of Age: NEPA's Lost Mandate, Duke Environmental Law & Policy Forum Vol. 21:113

Landau, D., Landau, S, Confidence Building Measures in Mediation, Mediation Quarterly, Vol. 15, no. 2 Winter 1997

Lomborg, Bjorn, Cool It The Skeptical Environmentalist's Guide to Global Warming, Random House, New York New York. (2007),

Reisner, M., Cadillac Desert: The American West and Its Disappearing Water, June 1, 1993

Roger, F. and Ury, W., Getting to Yes, Negotiating Agreements without Giving In, Second Edition Penguin Group, New York. (1991)

Mayer, B , The Dynamics of Conflict Resolution. San Francisco: Jossey-Bass, (2000).

Mayer, B), Beyond Neutrality, San Francisco: Jossey-Bass, (2004)

Meharg, Sara Measuring What Matters in Peace Operations and Crisis Management, Kingston ON, Canada, McGill-Queens University Press, (2009)

Odum, H.T., Environment, Power, and Society for the Twenty-First Century, The Hierarchy of Energy, Columbia University Press, (2007)

Plater Abrahams Graham, Hall Heinzerling Wirth, 539, Environmental Law and Policy: Nature, Law, and Society (1992)

Report to Congress on Recovery of Rare Earth Elements and Critical Minerals from Coal and Coal By-Products.pdf

Renner, M., Chafe, Z., (2007) Beyond Disasters; Creating Opportunities for Peace: Washington D.C., Worldwatch Institute.

Roger, F. and Ury, W., Getting to Yes, Negotiating Agreements without Giving In: 1997

RUHL, J.B. Water Wars, Eastern Style: Divvying Up the Apalachicola-Chattahoochee-Flint River Basin, 131 J. Contemp. Water Res. & Educ. (2005).

Ruhl, J.B., Equitable Apportionment of Ecosystem Services: New Water Law for a New Water Age, Vol. 19:1, Journal of Land Use, (2003),

Saleem, H. Ali Introduction: A Natural Connection between Ecology and Peace? Boston MA, MIT Press, (2007)

Sarat, Austin, Douglas, Lawrence. Umphrey, Martha, (2007) Law and Catastrophe, Standford, CA, Standford University Press

Sax & Abrams, Legal Control of Water Resources, West Publishing co. St. Paul Minn, 1986

Scheer, Goele (2008) Assessing Progress on the Road to Peace Planning, Monitoring and Evaluating Conflict Prevention and Peace Building, European Centre for Conflict Prevention/Global Secretariat of Global Partnership for the Prevention of Armed Conflict, Den Haag, Netherlands.

Shellenberger, M. and Nordhaus, T., (2004) The Death of Environmentalism: Global Warming Politics in a Post-Environmental World, www.Breakthrough.org

Simms, R.A., Equitable Apportionment-Priorities and Uses, Vol. 29, Natural Resources Journal, (1989),

Steady, Filomina, Environmental Justice in the New Millennium, New York, Macmillion, (2009)

Stewart, Cohen, Waddell, Melissa Climate Change in the 21st Century, McGill-Queens, London, (2009)

TOLUSE OLORUNNIPA & MICHAEL C. BENDER, Florida to Sue Georgia in U.S. Supreme Court Over Water (visited August 14, 2013), at http://www.bloomberg.com/news/2013-08-13/florida-to-suegeorgia-in-u-s-supreme-court-over-water.html; See also Florida's Motion for Leave to File a Complaint, Complaint, and Brief in Support of Motion (visited October 4, 2013), at http://www.eenews.net/assets/2013/10/02/document_gw_03.pdf.

Utton, A. Ein Search of an Integrating Principle for Interstate Water Law: Regulation versus the Market Place, Vol. 25, Natural Resource Journal, (1985)

Ury, W., Bret, J., Goldberg, S., getting Disputes Resolved: Designing Sustems to Cut the Costs of Conflict, (Jossey-Bass, London (1988)

U.S. Fid. & Guar. Co. v. Dick Corp./Barton Malow, 215 F.R.D. 503 (W.D. Pa. 2003).

WOLF, AARON T., Intl. Negotiation: A Journey of Theory and Practice, Vol. 5:2 (visited May 18, 2013), at http://www.transboundarywaters.orst.edu/publications/indigenous/.

www.ingramcontent.com/pod-product-compliance
Lightning Source LLC
LaVergne TN
LVHW022005200926
848481LV00007B/2238

* 9 7 9 8 2 9 5 7 3 2 6 5 2 *